AC Electrical Circuits Wc

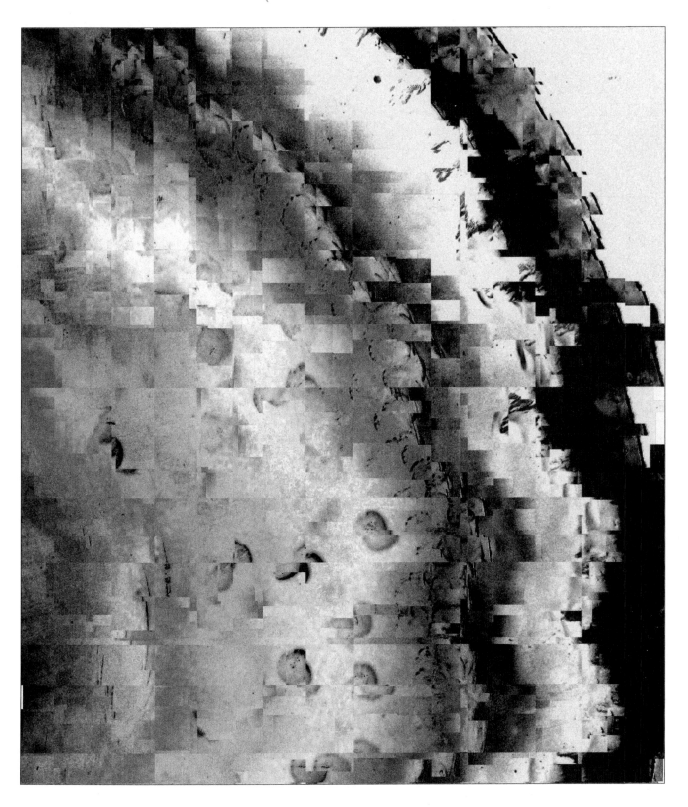

James M. Fiore

AC Electrical Circuits Workbook

by

James M. Fiore

Version 1.2.14, 25 November 2020

This **AC Electrical Circuits Workbook**, by James M. Fiore is copyrighted under the terms of a Creative Commons license:

This work is freely redistributable for non-commercial use, share-alike with attribution

Published by James M. Fiore via dissidents

ISBN13: 978-1796765083

For more information or feedback, contact:
James Fiore, Professor
Electrical Engineering Technology
Mohawk Valley Community College
1101 Sherman Drive
Utica, NY 13501
jfiore@mvcc.edu

For the latest revisions, related titles, and links to low cost print versions, go to:
www.mvcc.edu/jfiore or www.dissidents.com

YouTube Channel: *Electronics with Professor Fiore*

Cover art, *Some Thing Else*, by the author

Introduction

Welcome to the *AC Electrical Circuits Workbook*, an open educational resource (OER). The goal of this workbook is to provide a large number of problems and exercises in the area of AC electrical circuits to supplement or replace the exercises found in textbooks. It is offered free of charge under a Creative Commons non-commercial, share-alike with attribution license. This workbook has been replaced by the new text, *AC Electrical Circuit Analysis*, also OER. The new text features greatly expanded explanations and example material. As such, this workbook will no longer be updated.

If you are already familiar with the *DC Electrical Circuits Workbook*, the format of this title is similar. The workbook is split into several sections, each with an overview and review of the basic concepts and issues addressed in that section. These are followed by the exercises which are generally divided into four major types: analysis, design, challenge and simulation. Many SPICE-based circuit simulators are available, both free and commercial, that can be used with this workbook. The answers to most odd-numbered exercises can be found in the Appendix. A table of standard resistor sizes is also in the Appendix, which is useful for real-world design problems. If you have any questions regarding this workbook, or are interested in contributing to the project, do not hesitate to contact me.

This workbook is part of a series of OER titles in the areas of electricity, electronics, audio and computer programming. It includes five textbooks covering DC and AC circuit analysis, semiconductor devices, operational amplifiers, and embedded programming using the C language with the Arduino platform. There are seven laboratory manuals; one for each of the aforementioned texts plus computer programming using the Python language, and the science of sound. The most recent versions of all of my OER texts and manuals may be found at my MVCC web site as well as my mirror site: www.dissidents.com

This workbook was created using several free and open software applications including Open Office, Dia, and XnView.

"Everything should be made as simple as possible, but not simpler."

- Albert Einstein

Table of Contents

1 Fundamentals 7
Sinusoidal waveforms, basic Fourier analysis, complex numbers, reactance, impedance, susceptance, admittance, phasor diagrams.

2 Series RLC Circuits. 19
RLC circuits in series with either voltage sources or a current source.

3 Parallel RLC Circuits 34
RLC circuits in parallel with either a voltage source or current sources.

4 Series-Parallel RLC Circuits 45
RLC circuits using multiple components in series-parallel with either a single voltage source or current source.

5 Analysis Theorems and Techniques 57
Superposition theorem for multi-source circuits, source conversions, dependent sources, Thévenin's and Norton's theorems, maximum power transfer theorem, Pi-T (delta-Y) conversions.

6 Mesh and Nodal Analysis 82
Series-parallel RLC circuits using multiple voltage and/or current sources.

7 AC Power 101
Power waveforms, power triangle, power factor, power factor correction and efficiency.

8 Resonance 113
Series and parallel resonance.

9 Polyphase Power 128
Three phase systems in both delta and Y.

10 Magnetic Circuits and Transformers 138
Basic magnetic circuits using B-h curves, basic transformer operation.

Appendices

A: Standard Component Sizes 146
B: Answers to Selected Numbered Problems 147
C: Answers to Questions Not Asked 159

1 Fundamentals

This section covers:
- Sinusoidal waveforms.
- Basic Fourier analysis.
- Complex numbers.
- Reactance, impedance, susceptance and admittance.
- Phasor diagrams.

1.0 Introduction

Sinusoidal Waveforms

On the following page, Figure 1A is a representation of a sine wave, the simplest wave that may be created. It represents the displacement of a simple rotating vector (such as the second hand of a clock). Along the horizontal is the time axis. The vertical axis is represented here in general as a percentage of maximum but would ordinarily be a measurement of voltage, current, sound pressure, etc. Note the smooth variation that starts at zero, rises to a positive peak, falls back through zero to a negative peak, and then rises again to where it started. The whole process then repeats. Each repeat is referred to as a *cycle*. In the diagram, two complete cycles are shown. Sine waves exhibit *quarter wave symmetry*. That is, each quarter (in time) of the wave is identical to any other if you simply flip it around the horizontal axis and/or rotate it upright about its peak. The time it takes to complete one cycle is called the *period* and is denoted with the symbol T (for *T*ime). In the example below, the period is 10 milliseconds, or $T=10$ ms. The reciprocal of the period is the *frequency*, f. Thus, $f = 1/T$. The frequency indicates how many cycles exist in one second. To honor one of the 19th century researchers in the field, instead of calling the unit "cycles per second", we use *hertz*, named after Heinrich Hertz and abbreviated *Hz*. In the example above, $f = 1/10$ ms, or 100 Hz (100 cycles in one second).

The amplitude (vertical) of the wave can be expressed as a peak quantity, which would be the change from the center zero line up to the most positive value. Amplitude may also be expressed as peak-to-peak, the distance from the most negative to the most positive. Note that for a sine wave this will always be twice the peak value, although that may not be the case for other sorts of waves which may be asymmetrical. Alternately, amplitude may be given as an RMS (Root Mean Square) value. RMS is a special calculation used for finding equivalent DC power (very common, for example, with audio power amplifiers). For sine waves, RMS is always the peak value divided by the square root of two (approximately 1.414). As one over the square root of two is approximately 0.707, the RMS value of any sine wave is approximately 70.7 percent of its peak value. Again, this ratio would not necessarily be true of non-sine waves, and we will not concern ourselves with computing those other ratios. Finally, the ratio

7

of the peak value to the RMS value is called the *crest ratio*. This is a fixed value for sine waves (again, about 1.414), but can be over 10:1 for some kinds of audio signals.

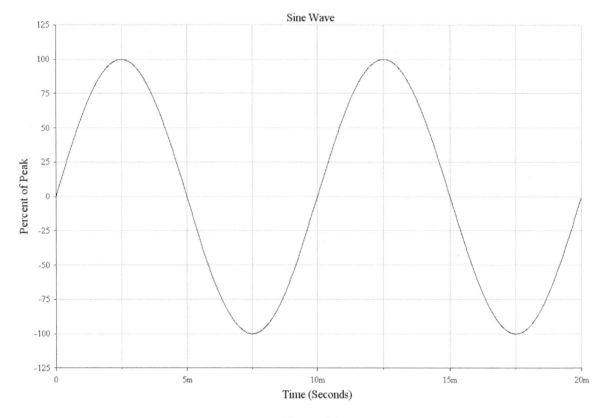

Figure 1A

Further, it is possible for a sine wave to be shifted in time compared to some other sine wave or reference. While it is possible to indicate this shift as an absolute time, it is more common to do so as a *phase shift*, that is, the time expressed as a portion of the period in degrees. For example, if one sine is ahead of another by one quarter of the period, it is said to be *leading* by 90° (i.e., ¼ of 360°). If it is behind by ½ of the period, it is said to be *lagging* by 180° (i.e., later in time by ½ cycle). AC waveforms may also be combined with a *DC offset*. Adding a positive DC level shifts the wave up vertically, and a negative DC level shifts the wave down vertically. This does not alter the frequency, phase or AC portion of the amplitude (although the absolute peaks would shift by the DC value).

Combining the foregoing elements allows us to develop a general format for a sine wave (voltage shown):

$$v(t) = V_{DC} + V_P \sin(2\pi f t + \theta)$$

Where
$v(t)$ is the voltage at some time t,
V_{DC} is the DC offset, if any,
V_P is the peak value,
f is the frequency,
θ is the phase shift (+ if leading and drawn to the left, − if lagging and drawn to the right).

Another item of interest is the speed of propagation of the wave. This varies widely. In the case of light in a vacuum (or to a close approximation, an electrical current in a wire), the velocity is about 3E8 meters per second (i.e., 300,000 km/s) or about 186,000 miles per second.

Given a velocity and a period, we can imagine how far apart the peaks of the wave are. This distance is called the *wavelength* and is denoted by the Greek letter lambda λ. Wavelength is equal to the velocity divided by the frequency, $\lambda = v/f$. Thus, for the 100 Hz waveform above, if this represents sound in air, $\lambda = 344$ m/s / 100 Hz, or 3.44 meters (a little over 11 feet). Notice that the higher the frequency, the shorter the wavelength. Also, note that the faster the velocity, the longer the wavelength.

Basic Fourier Analysis

The Fourier theorem, named after the French mathematician Jean-Baptiste Joseph Fourier, states that any repetitive waveform can be represented as a collection of sine and cosine waves of the proper amplitude and frequency. Alternately, it may be represented as a series of sine waves each with the proper amplitude, frequency and phase. This includes complex signals such as the human voice and musical instruments. Consequently, if a system is linear, by using superposition the response of a system to a complex wave may be understood in terms of its response to individual sine waves.

In this collection of waves, each component is known as a *partial* with the lowest frequency component known as the *fundamental*. All other partials are grouped together and referred to as *overtones*. "Regular" waveforms such as square waves and triangle waves feature a *harmonic overtone sequence* meaning that these overtones are integer multiples of the fundamental. As a shortcut, they are often referred to as just *harmonics*.

It might be hard to visualize initially, but like all complex waves, waves in the shape of a square or triangle must be made up of a series of sines. The general equation for a square wave is:

$$v(t) = \sum_{n=1}^{\infty} \frac{1}{2n-1} \sin((2n-1)2\pi f t)$$

This says that a square wave of frequency f is made up of an infinite series of sines at odd integer multiples of f, with an inverse amplitude characteristic. For example, a 100 Hz square consists of a 100 Hz sine plus a 300 Hz sine at 1/3 amplitude plus a 500 Hz sine at 1/5 amplitude plus a 700 Hz sine at 1/7 amplitude and so on.

A triangle wave is similar:

$$v(t) = \sum_{n=1}^{\infty} \frac{1}{(2n-1)^2} \cos((2n-1)2\pi ft)$$

Thus a triangle wave of frequency f is made up of an infinite series of cosines (sines with a 90 degree or one quarter cycle phase shift) at odd integer multiples of f, with an inverse square amplitude characteristic. For example, a 100 Hz triangle consists of a 100 Hz cosine plus a 300 Hz cosine at 1/9 amplitude plus a 500 Hz cosine at 1/25 amplitude plus a 700 Hz cosine at 1/49 amplitude and so on.

A series of graphs showing the construction of a square wave and a triangle wave follow. The square wave sequence begins with the fundamental and the first harmonic in Figure 1B. The result is an oddly bumpy wave. The second graph of Figure 1C adds the next two harmonics. As more harmonics are added, the sides get steeper and the top/bottom start to flatten. They flatten because each additional harmonic partially cancels some of the peaks and valleys from the previous summation. This gives rise to a greater number of undulations with each undulation being smaller in amplitude. The sequence finishes with Figure 1D showing seven harmonics being added with the result approaching a reasonable square wave. If more harmonics are added, the wave would approach a flat top and bottom with vertical sides.

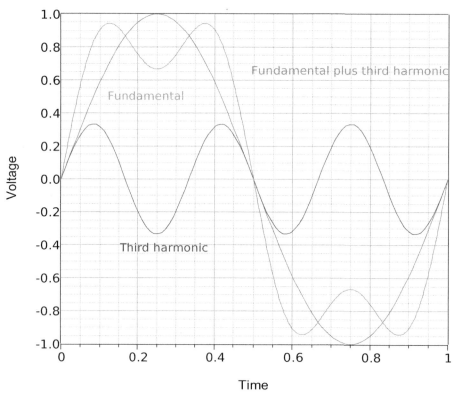

Figure 1B

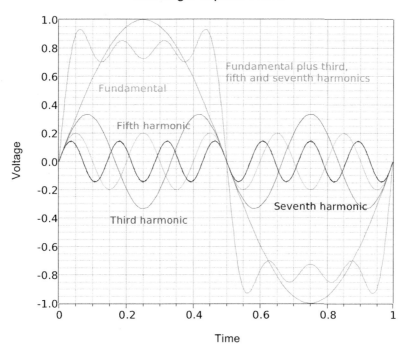

Figure 1C

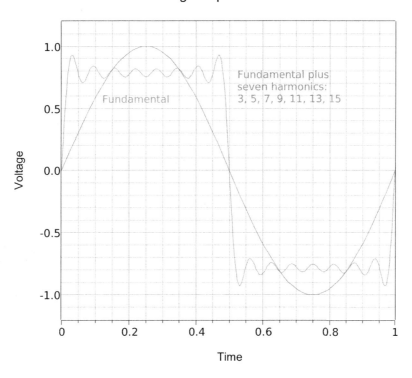

Figure 1D

The triangle sequence begins with a fundamental and the first harmonic as shown in Figure 1E. The resulting combination is already trending away from a simple sine shape. The second and final graph of Figure 1F shows a total of seven harmonics. The result is very close to a triangle, the only obvious deviation is the slight rounding at the very peaks. The addition of more harmonics would cause these to sharpen further.

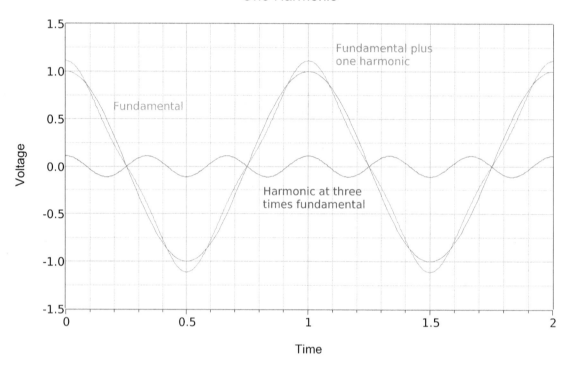

Figure 1E

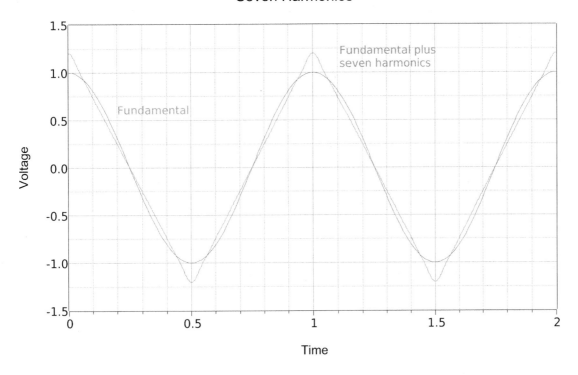

Figure 1F

Complex Numbers

In AC circuits, parameters such as voltage and current are complex numbers, that is, they have both a magnitude and a phase shift. For example, a voltage might be "12 volts at an angle of 30 degrees" (or more compactly, 12∠30). This is known as *polar form*. Alternately, a value can be broken into *rectangular form*, that is, right angle components consisting of a *real part* and an *imaginary part*. This can be visualized by imagining the horizontal and vertical components that define the point on an XY graph. The horizontal axis is the *real number* axis and the vertical axis is the *imaginary number* axis. The imaginary axis denotes values times the imaginary operator *j* (and often referred to as *i* outside of electrical analysis). The *j* operator is the square root of −1. An example would be 3+*j*4.

Converting from one form to another relies on basic trigonometric relations. If M is the polar magnitude with a phase of θ, and in rectangular form R is the real portion with an imaginary portion of I, then:

$$M = \sqrt{R^2 + I^2}$$
$$\theta = \tan^{-1}\frac{I}{R}$$
$$R = M \cos\theta$$
$$jI = M \sin\theta$$

To add or subtract complex quantities, first put them into rectangular form and then combine the reals with the reals and the j terms with the j terms as in $(3+j5) + (13-j1) = 16+j4$. These terms **must** be kept separate. $3+j5$ does **not** equal 8 (or even $j8$). That would be like saying that moving 3 feet to your right and 5 feet forward puts you in the same location as moving 8 feet to your right (or 8 feet forward).

The direct way to multiply or divide complex values is to first put them in polar form and then multiply or divide the magnitudes. The angles are added together for multiplication and subtracted for division. For example, $12\angle30$ times $2\angle45$ is $24\angle75$ while dividing them yields $6\angle-15$.

Reactance, Impedance, Susceptance and Admittance

Unlike a resistor, for an ideal capacitance or inductance, the voltage and current will not be in phase. For the ideal capacitor, the current leads the voltage across the capacitor by 90 degrees (the voltage across a capacitor cannot change instantaneously, $i=C\,dv/dt$). For an ideal inductor, the voltage leads the current by 90 degrees (the current through an inductor cannot change instantaneously, $v=L\,di/dt$). Consequently, if we divide the capacitor's voltage by its current (Ohm's law), we obtain a value with a phase angle of $-90°$. While the resultant is an ohmic value, it cannot be classified as a resistance. Instead, it is referred to as a *reactance* and denoted with the letter X. Thus we can refer to a capacitor's reactance, X_C, as some number at an angle of $-90°$, or more conveniently, we simply prepend $-j$, as in $X_C=-j75\ \Omega$. The case for the inductor is similar except that the angle is $+90$ degrees, and an example would be $X_L=j68\ \Omega$.

The reactance of an inductor is directly proportional to frequency:

$$X_L = +j\,2\pi f\,L$$

The reactance of a capacitor is inversely proportional to frequency:

$$X_C = -j\,\frac{1}{2\pi f\,C}$$

A mixture of resistance and reactance is known as *impedance*, Z. This can be visualized as a series combination of a resistor and either a capacitor or an inductor. Examples include $Z=100-j50\ \Omega$ (i.e., 100 ohms of resistance with 50 ohms of capacitive reactance) and $Z=600\angle45°\ \Omega$ (i.e., a magnitude of 600 ohms that includes resistance and inductive reactance).

Susceptance, B, is the reciprocal of reactance. *Admittance*, Y, is the reciprocal of impedance. These are similar to the relation between conductance and resistance, and are convenient for parallel circuit combinations.

Phasor Diagrams

A time domain representation of sine waves as drawn earlier tells us everything we need to know about the waves, however it is not the most compact method of displaying them. *Phasor diagrams* are used to show the relations of multiple waves. The diagram is based on a simple XY grid where the horizontal is the real axis and the vertical is the imaginary (*j*) axis. The magnitude and phase of each wave can then be drawn as a vector and the relationships between the waves is shown directly. For manual plotting, it is convenient to convert from polar form to rectangular form. In the phasor diagram of Figure 1G below, two vectors are shown: 8+*j*6 and 5−*j*3 (equivalent to 10∠36.9° and 5.83∠−31°). Phasor diagrams can be used to plot voltages, currents and impedances.

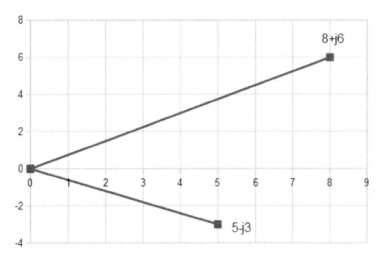

Figure 1G

1.1 Exercises

Analysis

1. Determine the AC peak and RMS voltages, DC offset, frequency, period and phase shift for the following expression: $v(t) = 10 \sin 2\pi\, 1000\, t$

2. Determine the AC peak and RMS voltages, DC offset, frequency, period and phase shift for the following expression: $v(t) = 0.4 \sin 2\pi\, 5000\, t$

3. Determine the peak AC portion voltage, DC offset, frequency, period and phase shift for the following expression: $v(t) = -3 + 20 \sin 2\pi\, 50\, t$

4. Determine the peak AC portion voltage, DC offset, frequency, period and phase shift for the following expression: $v(t) = 12 + 2 \sin 2\pi\, 20000\, t$

5. Determine the AC peak and RMS voltages, DC offset, frequency, period and phase shift for the following expression: $v(t) = 10 \sin (2\pi\, 100\, t + 45°)$

6. Determine the AC peak and RMS voltages, DC offset, frequency, period and phase shift for the following expression: $v(t) = 5 \sin (2\pi\, 1000\, t - 90°)$

7. Determine the peak AC portion voltage, DC offset, frequency, period and phase shift for the following expression: $v(t) = 10 + 1 \sin (2\pi\, 400\, t - 45°)$

8. Determine the peak AC portion voltage, DC offset, frequency, period and phase shift for the following expression: $v(t) = 10 + 10 \sin (2\pi\, 5000\, t + 30°)$

9. A 1 kHz sine wave has a phase of 72°. Determine the time delay. Repeat for a 20 kHz sine wave.

10. A 2 kHz sine wave has a phase of 18°. Determine the time delay. Repeat for a 100 kHz sine wave.

11. An oscilloscope measures a time delay of 0.2 milliseconds between a pair of 500 Hz sine waves. Determine the phase shift.

12. An oscilloscope measures a time delay of −10 microseconds between a pair of 20 kHz sine waves. Determine the phase shift.

13. Convert the following from rectangular to polar form: a) $10+j10$ b) $5-j10$ c) $-100+j20$ d) $3k+j4k$

14. Convert the following from rectangular to polar form: a) $2k+j1.5k$ b) $8-j8$ c) $-300+j300$ d) $-1k-j1k$

15. Convert these from polar to rectangular form: a) $10\angle 45°$ b) $0.4\angle 90°$ c) $-9\angle 60°$ d) $100\angle -45°$

16. Convert these from polar to rectangular form: a) $-4\angle 60°$ b) $-.9\angle 30°$ c) $5\angle 120°$ d) $6\angle -135°$

17. Perform the following computations: a) $(10+j10)+(5+j20)$ b) $(5+j2)+(-5+j2)$ c) $(80-j2)-(100+j2)$ d) $(-65+j50)-(5-j200)$

18. Perform the following computations: a) $(100+j200)+(75+j210)$ b) $(-35+j25)+(15+j8)$ c) $(500-j70)-(200+j30)$ d) $(-105+j540)-(5-j200)$

19. Perform the following computations: a) $(100+j200) \cdot (75+j210)$ b) $(-35+j25) \cdot (15+j8)$
    c) $(500 - j70) / (200+j30)$ d) $(-105+j540) / (5-j200)$

20. Perform the following computations: a) $(10+j10) \cdot (5+j20)$ b) $(5+j2) \cdot (-5+j2)$ c) $(80 - j2) / (100+j2)$
    d) $(-65+j50) / (5-j200)$

21. Perform the following computations: a) $(10\angle 0°) \cdot (10\angle 0°)$ b) $(5\angle 45°) \cdot (-2\angle 20°)$
    c) $(20\angle 135°) / (40\angle -10°)$ d) $(8\angle 0°) / (32\angle 45°)$

22. Perform the following computations: a) $(.3\angle 0°) \cdot (3\angle 180°)$ b) $(5\angle -45°) \cdot (-4\angle 20°)$
    c) $(.05\angle 95°) / (.04\angle -20°)$ d) $(500\angle 0°) / (60\angle 225°)$

23. Perform the following computations: a) $(.3\angle 0°) + (3\angle 180°)$ b) $(5\angle -45°) + (-4\angle 20°)$
    c) $(.05\angle 95°) - (.04\angle -20°)$ d) $(500\angle 0°) - (60\angle 225°)$

24. Perform the following computations: a) $(10\angle 0°) + (10\angle 0°)$ b) $(5\angle 45°) + (-2\angle 20°)$
    c) $(20\angle 135°) - (40\angle -10°)$ d) $(8\angle 0°) - (32\angle 45°)$

25. Determine the capacitive reactance of a 1 μF capacitor at the following frequencies: a) 10 Hz b) 500 Hz
    c) 10 kHz d) 400 kHz e) 10 MHz

26. Determine the capacitive reactance of a 220 pF capacitor at the following frequencies: a) 10 Hz
    b) 500 Hz   c) 10 kHz d) 400 kHz e) 10 MHz

27. Determine the capacitive reactance at 50 Hz for the following capacitors: a) 10 pF b) 470 pF c) 22 nF
    d) 33 μF

28. Determine the capacitive reactance at 1 MHz for the following capacitors: a) 22 pF b) 560 pF c) 33 nF
    d) 4.7 μF

29. Determine the inductive reactance of a 100 mH inductor at the following frequencies: a) 10 Hz
    b) 500 Hz c) 10 kHz d) 400 kHz e) 10 MHz

30. Determine the inductive reactance of a 220 μH inductor at the following frequencies: a) 10 Hz
    b) 500 Hz c) 10 kHz d) 400 kHz e) 10 MHz

31. Determine the inductive reactance at 1 kHz for the following inductors: a) 10 mH b) 500 mH
    c) 10 μH d) 400 μH

32. Determine the inductive reactance at 500 kHz for the following inductors: a) 1 mH b) 40 mH c) 2 μH
    d) 50 μH

33. Draw phasor diagrams for the following: a) $5+j2$ b) $-10-j20$ c) $8\angle 45°$ d) $2\angle -35°$

34. Draw phasor diagrams for the following: a) $60j-20$ b) $-40+j500$ c) $0.05\angle -45°$ d) $-15\angle 60°$

35. The fundamental of a certain square wave is a 5 volt peak, 1 kHz sine. Determine the amplitude and
    frequency of each of the next five harmonics.

36. The fundamental of a certain triangle wave is a 10 volt peak, 100 Hz sine. Determine the amplitude and
    frequency of each of the next five harmonics.

Design

37. Determine the capacitance required for the following reactance values at 1 kHz: a) 560 Ω b) 330 kΩ
    c) 470 kΩ d) 1.2 kΩ e) 750 Ω

38. Determine the capacitance required for the following reactance values at 20 Hz: a) 56 kΩ b) 330 kΩ
    c) 470 kΩ d) 1.2 kΩ e) 750 Ω

39. Determine the inductance required for the following reactance values at 100 MHz: a) 560 Ω b) 330 kΩ
    c) 470 kΩ d) 1.2 kΩ e) 750 Ω

40. Determine the inductance required for the following reactance values at 25 kHz: a) 56 Ω b) 33 kΩ
    c) 470 kΩ d) 1.2 kΩ e) 750 Ω

41. Which of the following have a reactance of less than 100 Ω for all frequencies below 1 kHz? a) 2 mH
    b) 99 mH c) 470 pF d) 10000 µF

42. Which of the following have a reactance of less than 8 Ω for all frequencies above 10 kHz? a) 10 nH
    b) 5 mH c) 56 pF d) 470 µF

43. Which of the following have a reactance of at least 1k Ω for all frequencies above 20 kHz? a) 2 mH
    b) 200 mH c) 680 pF d) 33 µF

44. Which of the following have a reactance of at least 75 Ω for all frequencies below 5 kHz? a) 680 µH
    b) 10 mH c) 82 pF d) 33 nF

Challenge

45. Determine the negative and positive peak voltages, RMS voltage, DC offset, frequency, period and phase
    shift for the following expression: $v(t) = -10 \sin (2\pi\ 250\ t + 180°)$

46. Determine the negative and positive peak voltages, DC offset, frequency, period and phase shift for the
    following expression: $v(t) = 1 - 100 \sin 2\pi\ 50000\ t$

47. Assume you have a DC coupled oscilloscope set as follows: time base = 100 microseconds/division,
    vertical sensitivity = 1 volt/division. Sketch the display of this waveform: $v(t) = 2 + 3 \sin 2\pi\ 2000\ t$

48. Assume you have a DC coupled oscilloscope set to the following: time base = 20 microseconds/division,
    vertical sensitivity = 200 millivolts/division. Sketch the display of this waveform:
    $v(t) = -0.2 + 0.4 \sin 2\pi\ 10000\ t$

49. A 200 Ω resistor is in series with a 1 mH inductor. Determine the impedance of this combination at 200
    Hz and at 20 kHz.

50. A 1 kΩ resistor is in series with an inductor. If the combined impedance at 10 kHz is 1.41 k∠45°,
    determine the inductance in mH.

2 Series RLC Circuits

This section covers:
- RLC circuits in series with either voltage sources or a current source.

2.0 Introduction

A series circuit is characterized by a single loop or path for current flow. Consequently, *the current is the same everywhere in a series circuit*. As resistances and reactances in series add, total resistance may be found by a vector summation of the individual components. Multiple voltage sources in series may also be added, however, polarities and phases must be considered as these potentials may partially cancel each other. In contrast, differing current sources are not placed in series as they would each attempt to establish a different series current, a practical impossibility.

Along with Ohm's law, the key law governing series circuits is Kirchhoff's voltage law, or KVL. This states that the sum of voltage rises and voltage drops around a series loop must equal zero. Alternately, it may be reworded as the sum of voltage rises around a series loop must equal the sum of voltage drops. As a pseudo formula: $\Sigma V\uparrow = \Sigma V\downarrow$. Note that due to phase shifts, summing *just the magnitudes* of the component voltages may produce a value much higher than the source voltage, appearing to violate KVL. KVL is **not** violated because phase shifts may produce partial cancellation. KVL will always work, but it **must** be performed using vector (phase sensitive) sums, **not** a scalar sum of magnitudes.

There are multiple techniques for solving series circuits. If all voltage source and component values are given, the circulating current can be found by dividing the equivalent voltage by the total impedance. Once the current is found, Ohm's law can be used to find the voltage drops across individual components. Alternately, the voltage divider rule may be used to find the voltage drops across the components(s) of interest. This rule exploits the fact that voltage drops in a series loop will be directly proportional to the size of the resistances/reactances. Thus, the voltage across any component resistance or reactance must equal the net supplied voltage times the ratio of the impedance of interest to the total impedance:

$$v_A = e \cdot Z_A/Z_{TOTAL}.$$

If the circuit uses a current source instead of a voltage source, then the circulating current is known and the voltage drop across any component may be determined directly using Ohm's law.

If the problem concerns determining resistance or reactance values, the basic idea will be to use these rules in reverse. For example, if a resistor value is needed to set a specific current, the total required impedance can be determined from this current and the given voltage supply. The values of the other series components can then be subtracted from the total (using rectangular form), yielding the required resistor or reactance value.

Similarly, if the voltages across two components are known, as long as one of the component values is known, the other can be determined using either the voltage divider rule or Ohm's law.

In sum, analysis for AC circuits follows that of DC circuit analysis; the key difference being the need to pay attention to the phase angles and performing vector computations. An example follows.

Consider the simple series circuit shown in Figure 2A. Assume the source is $10\angle 0°$, R is 4 kΩ and X_C is $-j3$ kΩ.

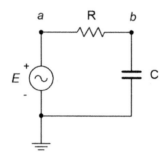

Figure 2A

The circuit impedance is 4 k $-j3$ k or 5 k$\angle -36.9°$ Ω. The circulating current is $10\angle 0°$ / 5 k$\angle -36.9°$, or 2E$-3\angle 36.9°$ amps. Using Ohm's law, v_R = 2E$-3\angle 36.9°$ · 4 k$\angle 0°$ = 8$\angle 36.9°$ volts. Also, v_C = 2E$-3\angle 36.9°$ · 3 k$\angle -90°$ = 6$\angle -53.1°$. In rectangular form, v_R = 6.4+j4.8 and v_C = 3.6$-$j4.8 volts. Note that summing these results in $10\angle 0°$, as expected. A phasor diagram of the voltages is shown below.

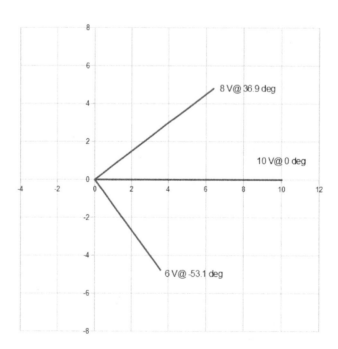

Figure 2B

20

2.1 Exercises

Analysis

1. Determine the impedance of the circuit of Figure 2.1 for a 1 kHz sine.

Figure 2.1

2. Determine the impedance of the circuit of Figure 2.1 for a 5 kHz sine.
3. Determine the impedance of the circuit of Figure 2.2 for a 10 kHz sine.

Figure 2.2

4. Determine the impedance of the circuit of Figure 2.2 for a 50 kHz sine.
5. Determine the impedance of the circuit of Figure 2.3 for a 1 kHz sine.

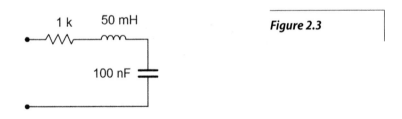

Figure 2.3

6. Determine the impedance of the circuit of Figure 2.3 for a 500 Hz sine.
7. Determine the impedance of the circuit of Figure 2.4.

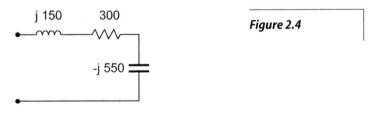

Figure 2.4

8. In the circuit of Figure 2.4, if the input frequency is 100 Hz, what is the value of the inductor, in mH?

9. In the circuit of Figure 2.4, if the input frequency is 200 Hz, what is the value of the capacitor, in µF?

10. Draw the voltage and current waveforms for the circuit of Figure 2.5.

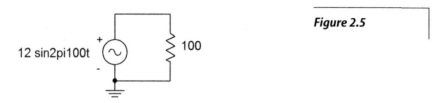

Figure 2.5

11. Draw the voltage and current waveforms for the circuit of Figure 2.6.

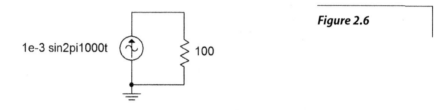

Figure 2.6

12. Draw the voltage and current waveforms for the circuit of Figure 2.7 if E is a one volt peak sine at a frequency of 10 kHz and C = 3.3 nF.

Figure 2.7

13. Draw the voltage and current waveforms for the circuit of Figure 2.8 if E is a two volt peak-peak sine at a frequency of 40 Hz and L = 33 mH.

Figure 2.8

14. Draw the voltage and current waveforms for the circuit of Figure 2.9 if I is a 10 micro-amp peak sine at a frequency of 2 kHz and C = 6.8 nF.

Figure 2.9

15. Draw the voltage and current waveforms for the circuit of Figure 2.10 if I is a two amp peak-peak sine at a frequency of 40 Hz and L = 33 mH.

Figure 2.10

16. Determine the impedance of the circuit of Figure 2.11.

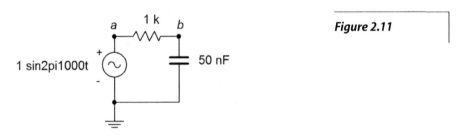

Figure 2.11

17. Determine the impedance of the circuit of Figure 2.11 using a frequency of 10 kHz.

18. For the circuit of Figure 2.11, determine the circulating current and the voltages across each component. Draw a phasor diagram of the three component voltages. Also find the time delay between the voltages of the components.

19. For the circuit of Figure 2.11 using a frequency of 10 kHz, determine the circulating current and the voltages across each component. Draw a phasor diagram of the three component voltages and determine the time delay between the capacitor and resistor voltages.

20. Determine the impedance of the circuit of Figure 2.12.

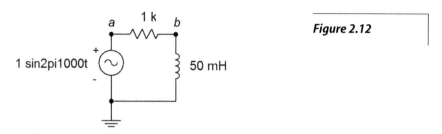

Figure 2.12

21. Determine the impedance of the circuit of Figure 2.12 using a frequency of 10 kHz.

22. For the circuit of Figure 2.12, determine the circulating current and the voltages across each component. Also find the time delay between the voltages of the components.

23. For the circuit of Figure 2.12 with a frequency of 3 kHz, determine the circulating current and the voltages across each component. Also find the time delay between the voltages of the components.

24. For the circuit of Figure 2.13, determine the circulating current.

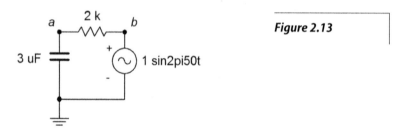

Figure 2.13

25. Determine the impedance of the circuit of Figure 2.13 using a frequency of 1.5 kHz.

26. For the circuit of Figure 2.13, determine the circulating current and the voltages across each component. Also find the time delay between the voltages of the components.

27. For the circuit of Figure 2.13 with a frequency of 1.5 kHz, determine the circulating current and the voltages across each component. Also find the time delay between the voltages of the components.

28. For the circuit of Figure 2.14, determine the circulating current and the voltages across each component. Draw a phasor diagram of the three component voltages and determine the time delay between the inductor and resistor voltages..

Figure 2.14

29. For the circuit of Figure 2.15, determine the circulating current and the voltages across each component.

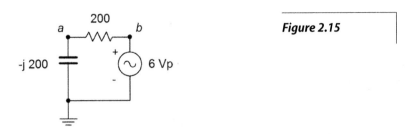

Figure 2.15

30. For the circuit of Figure 2.16, determine the circulating current and the voltages across each component.

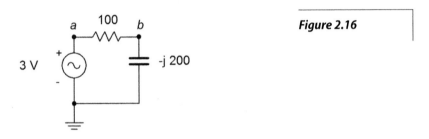

Figure 2.16

31. For the circuit of Figure 2.17, determine the applied voltage and the voltages across each component.

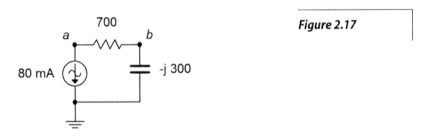

Figure 2.17

32. For the circuit of Figure 2.18, determine the applied voltage and the voltages across each component.

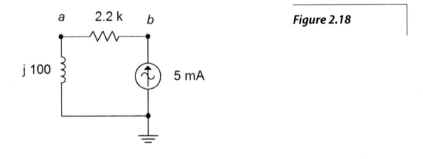

Figure 2.18

33. For the circuit of Figure 2.19, determine the circulating current and the voltages across each component.

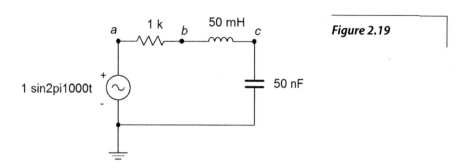

Figure 2.19

34. Repeat the previous problem using an input frequency of 10 kHz.

35. For the circuit of Figure 2.20, determine the circulating current and the voltages across each component. The source is a 10 volt peak sine at 20 kHz, R = 200 Ω, C = 100 nF and L = 1 mH.

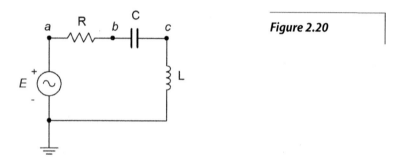

Figure 2.20

36. For the circuit of Figure 2.20, find v_b and v_{ac}.

37. For the circuit of Figure 2.21, find v_b and v_{ac}. The source is a 50 volt peak-peak sine at 10 kHz, R = 100 Ω, C = 200 nF and L = 1 mH.

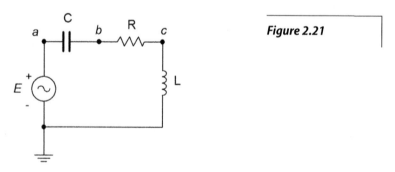

Figure 2.21

38. For the circuit of preceding problem, determine the circulating current and the voltages across each component.

39. For the circuit of Figure 2.22, determine the circulating current and the voltages across each component. E is a 1 volt peak 2 kHz sine. Also, draw a phasor diagram of the four component voltages.

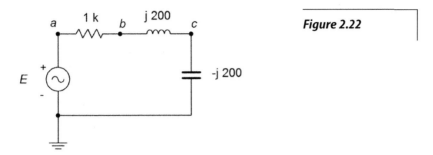

Figure 2.22

40. For the circuit of Figure 2.22, find v_b and v_{ca}. E is a 1 volt peak 2 kHz sine.

41. For the circuit of Figure 2.23, determine v_b, v_c and v_{ac}. E is a 10 volt peak 15 kHz sine.

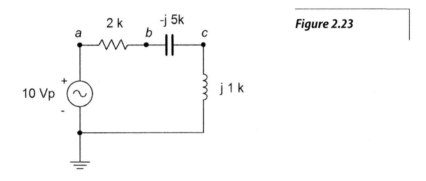

Figure 2.23

42. For the circuit of Figure 2.24, determine the circulating current and the voltages across each component. E is a 100 millivolt peak 250 Hz sine. Further, draw a phasor diagram of the four component voltages.

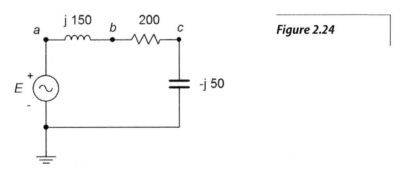

Figure 2.24

43. For the circuit of Figure 2.25, determine the circulating current and the voltages across each component. E is a 2 volt RMS 1 kHz sine. Also, draw a phasor diagram of the four component voltages.

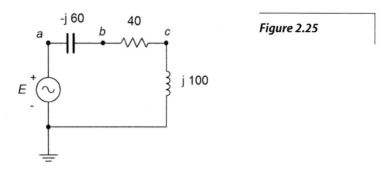

Figure 2.25

44. For the circuit of Figure 2.26, determine v_b, v_c and v_{ac}. E is a 1 volt peak 25 kHz sine.

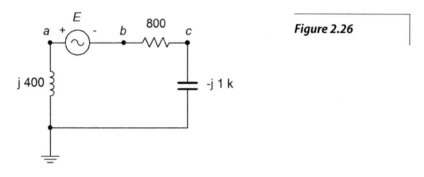

Figure 2.26

45. For the circuit of Figure 2.27, determine the voltages across each component. The source is a 50 mA peak sine at 15 kHz, R = 200 Ω, C = 100 nF and L = 1.5 mH.

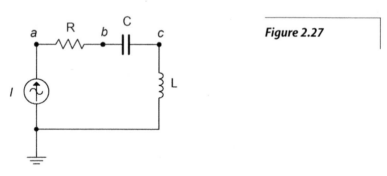

Figure 2.27

46. For the circuit of Figure 2.28, determine v_{ac}, v_b and v_c. The source is a 10 mA peak-peak sine at 50 kHz, R = 2 kΩ, C = 10 nF and L = 800 µH.

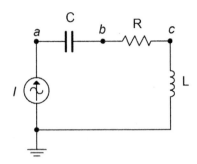

Figure 2.28

47. For the circuit of Figure 2.29, determine the voltages across each component. The source is a 2 mA RMS sine at 1 kHz, R = 1.2 kΩ, C = 750 nF and L = 6.8 mH.

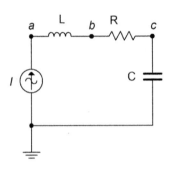

Figure 2.29

48. For the circuit of Figure 2.30, determine v_{ac}, v_b and v_a. The source is a 2 mA peak-peak sine at 300 kHz, R = 560 Ω, C = 6.8 nF and L = 400 µH.

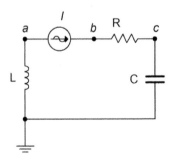

Figure 2.30

49. For the circuit of Figure 2.31, determine the voltages across each component.

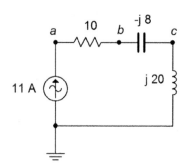

Figure 2.31

50. For the circuit of Figure 2.32, determine the voltages across each component. Further, draw a phasor diagram of the four component voltages.

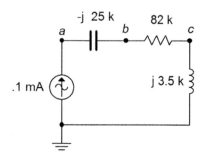

Figure 2.32

51. For the circuit of Figure 2.33, v_{ac}, v_b and v_c. The source is 5 mA peak at 8 kHz.

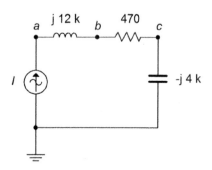

Figure 2.33

52. For the circuit of Figure 2.34, determine the voltages across each component. The source is 20 mA peak at 100 kHz.

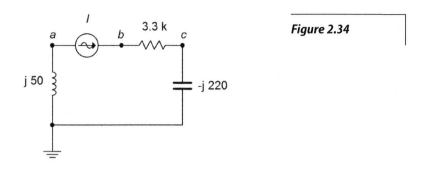

Figure 2.34

53. For the circuit of Figure 2.35, determine the voltages across each component.

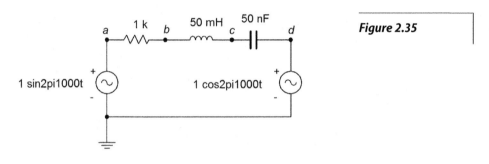

Figure 2.35

54. For the circuit of Figure 2.36, determine the voltages v_b and v_{db}. $E1=2\angle 0°$ and $E2=5\angle 90°$.

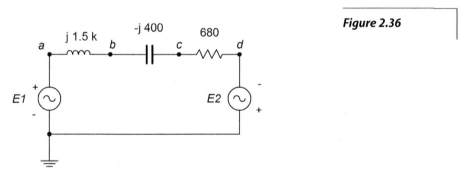

Figure 2.36

55. Determine the inductance and capacitance values for the circuit of problem 52.

56. For the circuit of Figure 2.36, determine the inductor and capacitor values if the source frequency is 12 kHz.

57. For the circuit of Figure 2.37, determine the voltages across each component. $E1=1\angle 0°$ and $E2=8\angle 60°$.

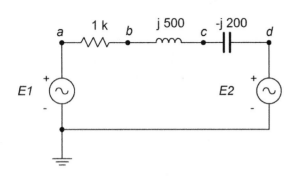

Figure 2.37

Design

58. Redesign the circuit of Figure 2.11 using a new capacitor such that the current magnitude from the source is 100 µA.

59. Redesign the circuit of Figure 2.12 using a new frequency such that the current magnitude from the source is 200 µA.

60. For the circuit of Figure 2.11, determine a new capacitor such that $|X_C| = R$.

61. For the circuit of Figure 2.12, determine a new frequency such that $|X_L| = R$.

Challenge

62. For the circuit of Figure 2.19, determine a new frequency such that $|X_C| = |X_L|$.

63. Determine the output voltage across the capacitor of Figure 2.11 at frequencies of 100 Hz, 5 kHz and 20 kHz. In light of this, if the input signal was a 1 kHz square wave instead of a sine wave as pictured, how would this circuit affect the shape of the output waveform (hint: consider superposition)?

64. Assume that you are troubleshooting a circuit like the one shown in Figure 2.20. E is a 2 volt peak sine at 2 kHz, R = 390 Ω, C = 100 nF and L = 25 mH. The circulating current measures approximately 4 mA with a lagging phase angle of just under −40 degrees. What is the likely problem?

65. Given the circuit shown in Figure 2.20, find the values for C and L if the source is a 6 volt sine wave at 1 kHz, R = 2 kΩ, v_R = 4 V and v_L = 5 V.

66. The circuit of Figure 2.38 can be used as part of a loudspeaker crossover network. The goal of this circuit is to steer the high frequency tones to the high frequency transducer (labeled here as "Loudspeaker" and often referred to as a *tweeter*). A similar network substitutes an inductor for the capacitor to steer the low frequency tones to the low frequency transducer (AKA *woofer*). These networks can be pictured as frequency sensitive voltage dividers. At very low frequencies, X_C is very large and blocks low frequency tones from reaching the tweeter. A mirror situation occurs with the

inductor/woofer variant. The *crossover frequency* is the frequency at which the reactance magnitude equals the resistance. Assuming simple 8 Ω resistances for the woofer and tweeter, determine capacitor and inductor values that would yield a 2.5 kHz crossover frequency. How might this concept be extended to a mid-range loudspeaker that only produces tones in the middle of the musical frequency spectrum? (Note, this concept will be revisited in the final simulation problem, below, and also in the Simulation portion of Part 4 which covers series-parallel circuits.)

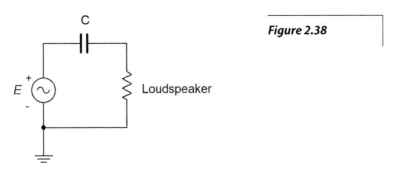

Figure 2.38

Simulation

67. Simulate the solution of design problem 58 and determine if the values produce the required results.

68. Simulate the solution of design problem 59 and determine if the values produce the required results.

69. Simulate the solution of design problem 60 and determine if the values produce the required results. Hint: if the reactance/resistance magnitudes are the same, then the voltage magnitudes will be identical.

70. Simulate the solution of design problem 61 and determine if the values produce the required results. Hint: if the reactance/resistance magnitudes are the same, then the voltage magnitudes will be identical.

71. Simulate the solution of challenge problem 62 and determine if the new frequency produces the required results. Hint: if the reactance magnitudes are the same, then the voltage magnitudes will be identical. Further, their phases will cause these voltages to cancel, leaving the resistor voltage equal to the source voltage.

72. Using a transient analysis, crosscheck the crossover design of the final challenge problem, above. Plot the resistor (loudspeaker) voltage across the range of 100 Hz to 20 kHz for both sections.

3 Parallel RLC Circuits

This section covers:
- RLC circuits in parallel with either a voltage source or current sources.

3.0 Introduction

Parallel circuits are in many ways the complement of series circuits. The most notable characteristic of a parallel circuit is that all components see the same voltage. Consequently, parallel circuits have only two nodes. Currents divide among the resistors in proportion to their conductance/susceptance (i.e., in inverse proportion to their resistance/reactance). Kirchhoff's current law (KCL) is the operative rule for parallel circuits. It states that the sum of all currents entering and exiting a node must sum to zero. Alternately, it can be stated as the sum of currents entering a node must equal the sum of currents exiting that node. As a pseudo formula:

$$\Sigma I \rightarrow = \Sigma I \leftarrow$$

It is possible to drive a parallel circuit with multiple current sources. These sources will add in much the same way that voltage sources in series add, that is, polarity and phase must be considered. Ordinarily, voltage sources with differing values are not placed in parallel as this violates the basic rule of parallel circuits (voltage being the same across all components).

Components in parallel combine as they do in DC circuits, again with the emphasis that phase must be included. The total admittance equals the sum of the admittances of the individual components,
$Y_T = Y_1 + Y_2 + Y_3 + ...$ or $1/Z_T = 1/Z_1 + 1/Z_2 + 1/Z_3 + ...$. Thus the equivalent impedance is equal to the reciprocal of the sum of the reciprocals, $Z_T = 1/(1/Z_1 + 1/Z_2 + 1/Z_3 + ...)$. For two components, the "product-sum rule" may be used, or $Z_T = (Z_1 \cdot Z_2)/(Z_1 + Z_2)$. Unlike the DC case of parallel resistors, the equivalent impedance of a parallel RLC circuit **does not** have to be smaller than the smallest component in that group. This is due to the phase angles of the reactive components.

The *current divider rule* remains valid for AC parallel circuits. Given two components, Z_1 and Z_2, and a current feeding them, I_T, the current through one of the components will equal the total current times the ratio of the *opposite* component over the sum of the impedance of the pair. For example, $i_1 = i_T \cdot Z_2/(Z_1 + Z_2)$. This rule is convenient in that the parallel equivalent impedance need not be computed, but remember, *it is valid only when there are just two components involved.*

When analyzing a parallel circuit, if it is being driven by a voltage source, then this same voltage must appear across each of the individual components. Ohm's law can then be used to determine the individual currents. According to KCL, the total current exiting the source must be equal to the sum of these individual currents. For example, in the circuit shown in Figure 3A, the voltage E must appear across both R and L. Therefore, the currents must be $i_L = E/X_L$ and $i_R = E/R$, and $i_{total} = i_L + i_R$. i_{total} can also be found be determining the parallel

34

equivalent impedance of R and L, and then dividing this into *E*. This technique can also be used in reverse in order to determine a resistance or reactance value that will produce a given total current: dividing the source by the current yields the equivalent parallel impedance. As one of the two is already known, the known component can be used to determine the value of the unknown component.

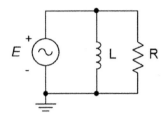

Figure 3A

To demonstrate, assume in the circuit of Figure 3A that the source is $10\angle 0°$, $X_L = j2$ kΩ and $R=1$ kΩ. The resulting currents would be $i_L = 5\text{E}-3\angle -90°$ amps, $I_R = 10\text{E}-3\angle 0°$ amps, $I_{total} = 11.2\text{E}-3\angle -26.6°$ amps. This current can also be found by dividing the source voltage by the total impedance. The equivalent parallel impedance is $894\angle 26.6°$ Ω. Interestingly, in rectangular form, this is $800+j400$, meaning that this parallel combination is equivalent to a series combination of an 800 ohm resistance and a 400 ohm inductive reactance. A phasor diagram of the currents is shown in Figure 3B, below.

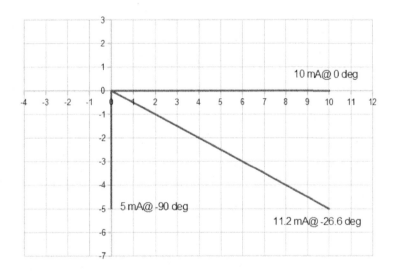

Figure 3B

If the parallel circuit is driven by a current source, as shown in Figure 3C, there are two basic methods of solving for the component currents. The fastest method is to simply use the current divider rule. If desired, the component voltage can then be found using Ohm's law. An alternate method involves finding the parallel equivalent impedance first, and then using Ohm's law to determine the voltage (remember, being a parallel circuit, there is only one common voltage). Given the voltage, Ohm's law can be used to find the current through one component. To find the current through the other, either Ohm's law can be applied a second time, or KCL may be used, subtracting the current through the first component from the source current. If there are more than two components, usually the second method would be the most efficient course of action.

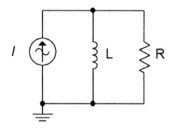

Figure 3C

It is worth noting that both methods described above will yield the correct answers. One is not "more correct" than the other. We can consider each of these as a separate "solution path"; that is, a method of arriving at the desired end point. In general, the more complex the circuit, the more solution paths there will be. This is good because one path may be more obvious to you than another. It also allows you a means of cross-checking your work.

3.1 Exercises

Analysis

1. Determine the effective impedance of the network shown in Figure 3.1 at 10 MHz.

   Figure 3.1

2. Determine the effective impedance of the network shown in Figure 3.2 at 100 Hz.

   Figure 3.2

3. Determine the effective impedance of the network shown in Figure 3.3 at 5 kHz.

Figure 3.3

4. Determine the effective impedance of the network shown in Figure 3.4 at 20 kHz.

Figure 3.4

5. Determine the effective impedance of the network shown in Figure 3.5.

Figure 3.5

6. Determine the effective impedance of the network shown in Figure 3.5 if the frequency is halved and if the frequency is doubled.

7. For the network shown in Figure 3.1, determine the frequency below which the impedance is mostly resistive.

8. For the network shown in Figure 3.2, determine the frequency below which the impedance is mostly inductive.

9. Draw phasor impedance plot for problem 1.

10. Draw phasor impedance plot for problem 2.

11. Determine the three branch currents for the circuit shown in Figure 3.6 and draw their phasor diagram.

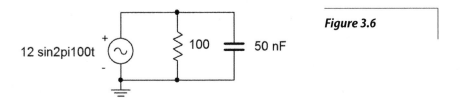

Figure 3.6

12. Determine the three branch currents for the circuit shown in Figure 3.7 and draw their phasor diagram.

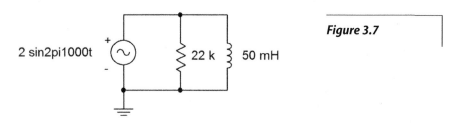

Figure 3.7

13. Determine the four branch currents for the circuit shown in Figure 3.8 and draw their phasor diagram.

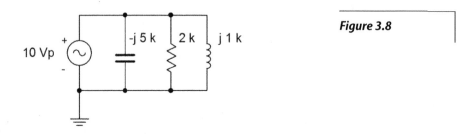

Figure 3.8

14. Determine all of the branch currents for the circuit shown in Figure 3.9 assuming E is a 1 volt RMS sine.

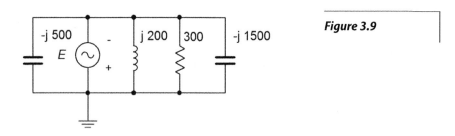

Figure 3.9

15. Determine all of the branch currents for the circuit shown in Figure 3.10 given $E = 10$ volt peak sine, $R = 220$, $X_C = -j\,500$. and $X_L = j1.5$ k.

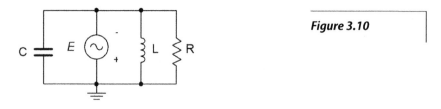

Figure 3.10

16. Determine all of the branch currents for the circuit shown in Figure 3.11 given $E = 2$ volt peak sine, $R = 1$ k, $X_C = -j2$ k. and $X_L = j3$ k.

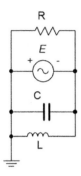

Figure 3.11

17. Determine the component currents for the circuit shown in Figure 3.12. Draw phasor diagram of the source and branch currents.

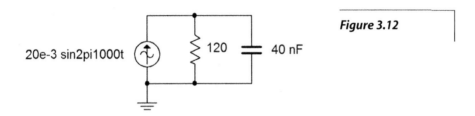

Figure 3.12

18. Determine the resistor and capacitor voltages for the circuit shown in Figure 3.12.

19. Determine the resistor and inductor voltages for the circuit shown in Figure 3.13.

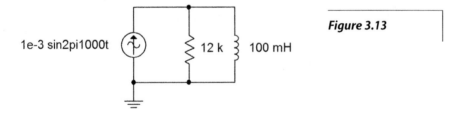

Figure 3.13

20. Determine the component currents for the circuit shown in Figure 3.13. Draw phasor diagram of the source and branch currents.

21. Determine the source voltage for the circuit shown in Figure 3.14.

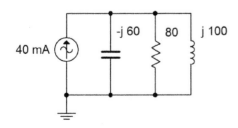

Figure 3.14

22. Determine the component currents for the circuit shown in Figure 3.14. Draw phasor diagram of the source and branch currents.

23. Determine the component currents for the circuit shown in Figure 3.15. *I* is 20 mA at 0 degrees.

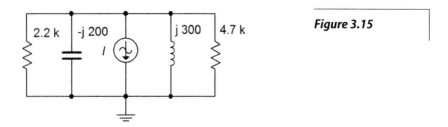

Figure 3.15

24. Determine the source voltage for the circuit shown in Figure 3.15. *I* is 20 mA at 0 degrees.

25. Determine the source voltage for the circuit shown in Figure 3.16. Assume *I1* is 1 mA at 0 degrees and *I2* is 2 mA at +90 degrees.

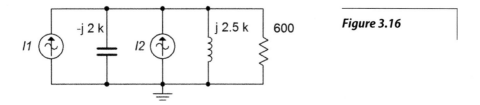

Figure 3.16

26. Determine the capacitor and inductor currents in the circuit of Figure 3.16. Assume *I1* is 1 mA at 0 degrees and *I2* is 2 mA at +90 degrees.

27. Determine the resistor and capacitor currents in the circuit of Figure 3.17. Assume *I1* is 2 A at 0 degrees and *I2* is 500 mA at +45 degrees.

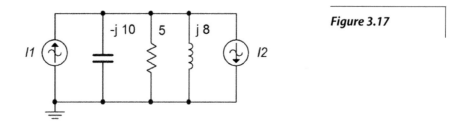

Figure 3.17

28. Determine the source voltage for the circuit shown in Figure 3.17. Assume *I1* is 2 A at 0 degrees and *I2* is 500 mA at +45 degrees.

Design

29. For the network shown in Figure 3.18, determine a value of C such that the impedance magnitude of the circuit is 1 kΩ. The source is a 50 Hz sine and R is 2.2 kΩ.

Figure 3.18

30. For the network shown in Figure 3.19, determine a value of L such that the impedance magnitude of the circuit is 2 kΩ. The source is a 2 MHz sine and R is 3.3 kΩ.

Figure 3.19

31. For the circuit shown in Figure 3.18, determine a value for C such that the magnitude of the source current is 1 mA. E is a 2 volt 10 kHz sine and R = 8 kΩ.

32. For the network shown in Figure 3.19, determine a value for L such that the magnitude of the source current is 10 mA. E is a 25 volt 100 kHz sine and R = 4 kΩ.

33. For the network shown in Figure 3.20, determine a value of C such that the impedance magnitude of the circuit is 10 kΩ. The source is a 440 Hz sine and R is 33 kΩ.

Figure 3.20

34. For the network shown in Figure 3.21, determine a value of L such that the impedance magnitude of the circuit is 200 Ω. The source is a 60 Hz sine and R is 680 Ω.

Figure 3.21

35. For the circuit shown in Figure 3.20, determine a value for C such that the magnitude of the circuit voltage is 200 volts. The source current is a 100 mA 1200 Hz sine and R = 15 kΩ.

36. For the circuit shown in Figure 3.21, determine a value for L such that the magnitude of the circuit voltage is 50 volts. The source current is a 2.3 A 60 Hz sine and R = 330 Ω.

37. Given the circuit shown in Figure 3.18, determine a value for C such that the impedance angle is −45 degrees. The source a 1 volt peak sine at 600 Hz and R = 680 Ω.

38. Given the circuit shown in Figure 3.19, determine a value for L such that the impedance angle is 45 degrees. The source a 10 volt peak sine at 100 kHz and R = 1.2 kΩ.

39. Determine a value for C such $|X_C| = |X_L|$ for the circuit shown in Figure 3.22. The source frequency is 1 kHz, R = 200 Ω and L = 50 mH.

Figure 3.22

40. Determine a value for L such $|X_C| = |X_L|$ for the circuit shown in Figure 3.23. The source frequency is 22 kHz, R = 18 kΩ and C = 5 nF.

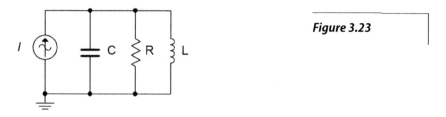

Figure 3.23

41. Add one or more components in parallel with the circuit of Figure 3.2 such that the resulting impedance at 20 Hz is 10 Ω with a phase angle of at least +30°.

Challenge

42. Determine a value for C such that the impedance angle for the circuit shown in Figure 3.22 is purely resistive (0 degrees). The source frequency is 1 kHz, R = 200 Ω and L = 50 mH.

43. Is it possible to change the value of the resistor in Figure 3.14 so that the system voltage is 4 volts? If so, what is the value? If not, why not?

44. Is it possible to change the value of the inductor and/or capacitor in Figure 3.14 so that the system voltage is 4 volts? If so, what is/are the values? If not, why not?

45. Assume you are troubleshooting a circuit like the one shown in Figure 3.23. I is a 10 mA peak sine at 2 kHz, R = 390 Ω, C = 200 nF and L = 25 mH. The measured resistor voltage is a little under 2.5 volts. What is the likely culprit?

46. Given the circuit shown in Figure 3.23, find the values for C and L if the source is a sine wave at 1 kHz, R = 4 kΩ, i_{Source} = 3 mA, i_R = 2 mA, i_L = 5 mA,

Simulation

47. Using a transient analysis simulation, verify that the source current magnitude is 1 mA using the capacitor value determined in design problem 31.

48. Using a transient analysis simulation, verify that the source current magnitude is 10 mA using the inductor value determined in design problem 32.

49. Using a transient analysis simulation, verify that the source voltage magnitude is 200 volts using the capacitor value determined in design problem 35.

50. Using a transient analysis simulation, verify that the source voltage magnitude is 50 volts using the inductor value determined in design problem 36.

51. Using a transient analysis simulation, verify the design solution for problem 39. This can be checked by seeing if the current magnitudes in C and L are identical.

52. Using a transient analysis simulation, verify the design solution for problem 40. This can be checked by seeing if the current magnitudes in C and L are identical.

53. Impedance magnitude as a function of frequency can be investigated by driving the circuit with a fixed amplitude current source across a range of frequencies. The resulting voltage will be proportional to the effective impedance. Investigate this effect by performing an AC analysis on the circuits shown in Figures 3.12 and 3.13. Use a frequency range of 10 Hz to 1 MHz. Before running the simulations, sketch your expected results.

54. Following the idea presented in the previous problem, investigate the impedance as a function of frequency of the circuit shown in Figure 3.23. Use R = 1 k, C = 10 nF, and L = 1 mH. Run the simulation from 100 Hz to 10 MHz. Make sure to sketch your expected results first.

Notes

4 Series-Parallel RLC Circuits

This section covers:
- RLC circuits using multiple components in series-parallel with either a single effective voltage or current source.

4.0 Introduction

This section deals with a subset of series-parallel RLC circuits, specifically those that are driven by a single effective current or voltage source, and which may be simplified using series and parallel component combinations. The key to analyzing series-parallel circuits is in recognizing portions of the circuit that are in series or in parallel and then applying the series and parallel analysis rules to those sections. Ohm's law, KVL and KCL may be used in turn to solve portions of the problem until all currents and voltages are found. As individual voltages and currents are determined, this makes it easier to apply these rules to determine other values.

Consider the circuit of Figure 4A. This is neither just a series circuit nor just a parallel circuit. If it was a series circuit then the current through all components would have to be same, that is, there would no nodes where the current could divide. This is clearly not the case as the current flowing through the capacitor can divide at node b, with one portion flowing down through the resistor and the remainder through the inductor. On the other hand, if it was strictly parallel, then all of the components would have to exhibit the same voltage and therefore there would be only two connection points in the circuit. This is also not the case as there are three such points: a, b and ground.

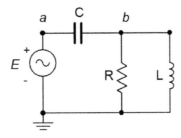

Figure 4A

What is true for this circuit is that the resistor and the inductor are in parallel. We know this because both components are attached to the same two nodes, b and ground, and must exhibit the same voltage, v_b. As such, we can find the equivalent impedance of this pair and treat the result as a single value, let's call it Z_x. In this newly simplified circuit, Z_x is in series with the capacitor and the source. We have simplified the original circuit into a series circuit and thus the series circuit analysis rules may be applied.

There are many solution paths at this point. For example, we could find the total impedance, Z_t, by adding Z_x to X_c. Dividing this by the source voltage yields the total current flowing out of the source, i_{total}. This current must flow through the capacitor so Ohm's law can be used to find the voltage drop across it. This same current must be flowing through Z_x, so Ohm's law can be used to find the associated voltage (v_b). The currents through the parallel resistor/inductor combo may then be found using Ohm's law for each element (e.g., the current through R must be v_b/R). Alternately, these currents may be found by using the current divider rule (e.g., the current through R must be $i_{total} \cdot X_L/Z_x$; remember, the current divider rule uses the ratio of the *opposite* component over the sum).

Another solution path would be to apply the voltage divider rule to X_c and Z_x in order to derive the two voltage drops (or the rule can be applied to find just one of the drops and the other voltage may be found by subtracting that from the source, an application of KVL). Once the voltages are determined, Ohm's law can be used to find the currents. Much like the analysis of DC series-parallel circuits, the larger the circuit becomes, the greater the number of possible solution paths. It is often useful to solve these circuits using a variety of techniques as a means of cross-checking the results and sharpening the skill set.

As an example, consider the solution of the circuit shown above in Figure 4A using a 10 volt source, $-j200$ for the capacitive reactance, $j300$ for the inductive reactance and a resistance of 100 Ω. The parallel resistor/inductor combo, Z_x, is $94.9\angle18.4°$ Ω, or $90+j30$. Placing this in series with $-j200$ yields a total impedance, Z_{total}, of $192.4\angle-62.1°$ Ω, or $90-j170$. Using the voltage divider rule to find v_b via the source voltage of $10\angle0°$ yields $4.93\angle80.5°$ volts, or $0.814+j4.86$. The divider rule can be used a second time to find the capacitor voltage, $10.4\angle-27.9°$ volts, or $9.188-j4.86$.

Note that the capacitor voltage is *greater than* the source. This is not something that would happen with a simple resistive circuit. Equally important, note that adding these two potentials results in the source voltage of $10\angle0°$, as expected. A phasor diagram of these voltages is shown in Figure 4B on the following page.

Finally, it is worth mentioning that the phase angle between the two main portions is not 90 degrees as is often the case in simpler series-only and parallel-only circuits (see the phasor plots in the two prior sections for illustration). This is due to the fact that the subsections themselves are not just resistance or reactance, but rather, a complex impedance (consider for example Z_x). If the component vectors are connected head-to-tail graphically, the result would yield the source voltage, as expected. KVL cannot be violated.

Series-parallel simplification techniques will not work for all circuits. Some networks such as delta or bridge configurations require other techniques that will be addressed in later sections.

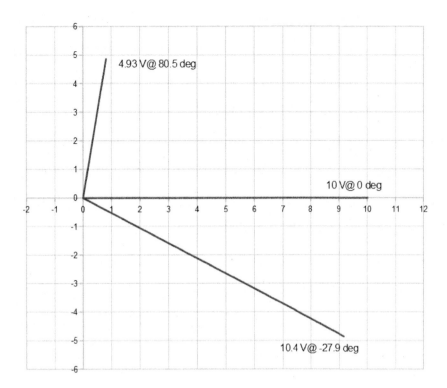

Figure 4B

4.1 Exercises

Analysis

1. Determine the impedance of the circuit of Figure 4.1 at frequencies of 100 Hz, 10 kHz and 1 MHz.

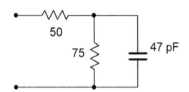

Figure 4.1

2. Determine the impedance of the circuit of Figure 4.2 at frequencies of 20 Hz, 1 kHz and 20 kHz.

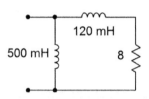

Figure 4.2

3. Determine the impedance of the circuit of Figure 4.3 at frequencies of 300 Hz, 30 kHz and 3 MHz.

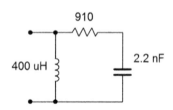

Figure 4.3

4. Determine the impedance of the circuit of Figure 4.4 at frequencies of 1 kHz, 20 kHz and 1 MHz.

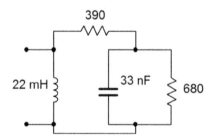

Figure 4.4

5. Determine the impedance of the circuit of Figure 4.5.

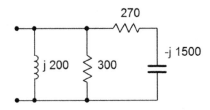

Figure 4.5

6. Determine the impedance of the circuit of Figure 4.6.

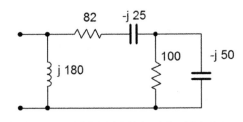

Figure 4.6

7. For the circuit of Figure 4.7, determine the source current and the current through each of the components.

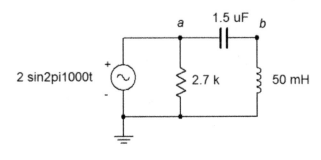

Figure 4.7

8. For the circuit of Figure 4.7, determine voltages v_{ab} and v_b.

9. For the circuit of Figure 4.8, determine voltages across R, L and C if the source is 7 volts RMS.

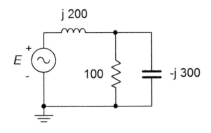

Figure 4.8

10. For the circuit of Figure 4.8, determine the source current and the current through each of the three components. Also, draw a phasor diagram of E, v_L and v_R.

11. For the circuit of Figure 4.9, determine the source current and the current through each of the components.

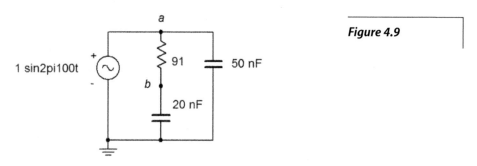

Figure 4.9

12. For the circuit of Figure 4.9, determine voltages v_{ab} and v_b. Also, draw a phasor diagram of E, v_{ab} and v_b.

13. For the circuit of Figure 4.10, determine voltages v_{ab} and v_b if the source is 20 volts peak.

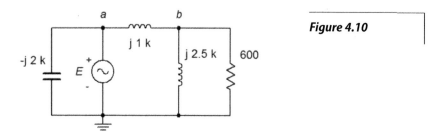

Figure 4.10

14. For the circuit of Figure 4.10, determine the source current and the current through each of the four components if the source is 20 volts peak.

15. For the circuit of Figure 4.11, determine the source current and the current through each of the four components.

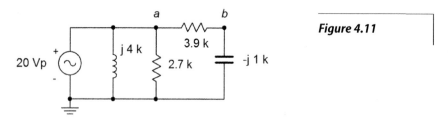

Figure 4.11

16. For the circuit of Figure 4.11, determine voltages v_{ab} and v_b.

17. For the circuit of Figure 4.12, determine voltages v_{ab} and v_b if the source is 100 volts peak.

18. For the circuit of Figure 4.12, determine the currents through the two resistors.

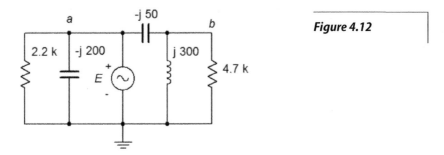

Figure 4.12

19. For the circuit of Figure 4.13, determine the currents each of the three components.

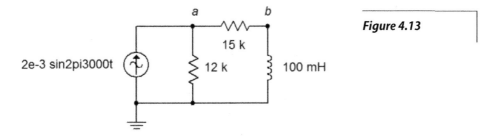

Figure 4.13

20. For the circuit of Figure 4.13, determine voltages v_a and v_b.

21. For the circuit of Figure 4.14, determine voltages v_a and v_b.

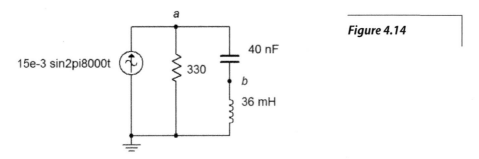

Figure 4.14

22. For the circuit of Figure 4.14, determine the middle and right branch currents and draw a phasor diagram of three circuit currents.

51

23. For the circuit of Figure 4.15, determine voltages v_a and v_b.

24. For the circuit of Figure 4.15, determine the currents through the two resistors.

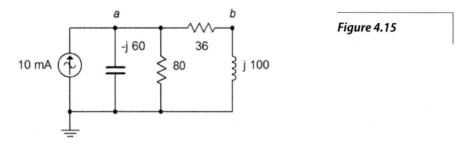

Figure 4.15

25. For the circuit of Figure 4.16, determine voltages v_a and v_b. $I = 25$ mA.

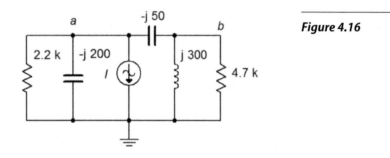

Figure 4.16

26. For the circuit of Figure 4.16, determine the currents through the two capacitors.

27. For the circuit of Figure 4.17, determine the current through the capacitor. $I1 = 10\text{E}{-}3\angle 0°$ A and $I2 = 3\text{E}{-}3\angle 90°$ A.

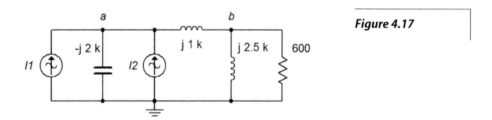

Figure 4.17

28. For the circuit of Figure 4.17, determine voltages v_a and v_b. $I1 = 10\text{E}{-}3\angle 0°$ A and $I2 = 3\text{E}{-}3\angle 90°$ A.

29. For the circuit of Figure 4.18, determine voltages v_a and v_b. $I1 = 2\angle 45°$ A and $I2 = 0.5\angle 0°$ A.

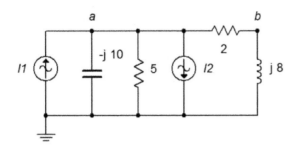

Figure 4.18

30. For the circuit of Figure 4.18, determine the currents through the two resistors. $I1 = 2\angle 45°$ A and $I2 = 0.5\angle 0°$ A.

31. For the bridge circuit of Figure 4.19, determine v_{ab}. The source is 50 volts peak.

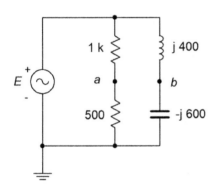

Figure 4.19

32. For the bridge circuit of Figure 4.20, determine v_{ab}. The source is 6 amps peak.

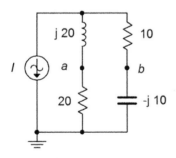

Figure 4.20

Design

33. Determine a new value for the capacitor in Figure 4.7 such that v_b is 1.5 volts.

34. Determine the required inductive reactance in Figure 4.8 to shift the capacitor voltage to half of the source voltage.

35. Determine a new value for the 20 nF capacitor in Figure 4.9 such that the resistor current is 2 mA.

36. In the circuit of Figure 4.21, determine a value for L such that the magnitude of v_b equals $v_a/2$ if the source frequency is 10 kHz, R = 2.7 kΩ and C = 10 nF.

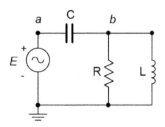

Figure 4.21

37. Given the circuit of Figure 4.21, determine a value for C such that the source current is in phase with the source voltage. The source frequency is 1 kHz, R = 68 Ω and L = 22 mH.

38. Given the circuit of Figure 4.22, determine a value for L such that v_b is 1 volt. The source is a 6 volt peak sine at 50 kHz, R1 = 510 Ω and R2 = 220 Ω.

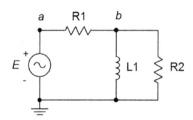

Figure 4.22

39. Given the circuit of Figure 4.19, determine a new value for the inductor such that the magnitude of v_b equals the magnitude of v_a. Assume that the source frequency is 20 kHz.

Challenge

40. Consider the circuit drawn in Figure 4.23. Using only the available components of 1 kΩ, 2.2 kΩ, 1 mH, 5 mH, 10 nF, 75 nF and 560 nF, is it possible to configure a circuit such that v_a is half the magnitude of v_b? If so, indicate which values could be used for the four components. If not, explain your reasoning.

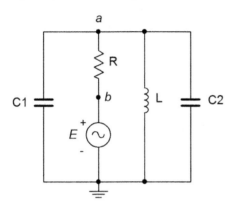

Figure 4.23

41. Given the circuit of Figure 4.9, determine the frequency at which v_b is half of the source voltage.

42. For the circuit of Figure 4.24, determine voltages v_a, v_b, and v_c. $I1 = 5\angle 0°$ A and $I2 = 3\angle 90°$ A.

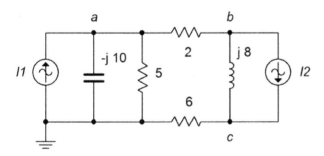

Figure 4.24

43. Given the circuit of Figure 4.19, is it possible to change the values of the two resistors such that the phase angle of v_a is the same as that of v_b? If so, what are the new values, and if not, explain why it is not possible.

Simulation

44. Perform a transient analysis to verify the node voltages computed for problem 8.

45. Perform a transient analysis to verify the node voltages computed for problem 12.

46. Perform a transient analysis to verify the node voltages computed for problem 20.

47. Perform a transient analysis to verify the node voltages computed for problem 21.

48. Consider the circuit of problem 17. Assuming the source frequency is 10 kHz, determine values for the capacitors and inductors. Then, use a transient analysis to verify the results of problem 17.

49. Perform a transient analysis on the result of problem 33 to verify the accuracy of the design.

50. Perform a transient analysis on the result of problem 34 to verify the accuracy of the design.

51. Use an AC frequency response simulation to verify the results of problem 41.

52. The concept of a loudspeaker crossover network was presented in Part 2 Series RLC Circuits. By adding more components, it is possible to increase the rate of attenuation so that the undesired signals are further reduced in amplitude. The circuits of Figure 4.25 and 4.26 show crossovers for the woofer and tweeter, respectively. Assuming standard 8 Ω loudspeakers, use an AC frequency domain simulation to determine the crossover frequency of each network. Also, compare the curves at node *a* to those at node *b*. Finally, compare the attenuation slopes to those generated by the simpler crossover network presented at the end of Part 2. Component values for the woofer: L1 = 760 µH, L2 = 250 µH, C = 10.6 µF. Component values for the tweeter: C1 = 5.3 µF, C2 = 16 µF, L = 380 µH.

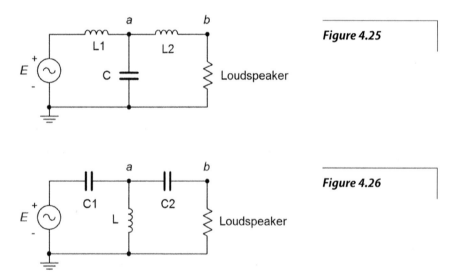

Figure 4.25

Figure 4.26

5 Analysis Theorems and Techniques

This section covers:
- Superposition theorem for multi-source circuits.
- Source conversions.
- Dependent sources.
- Thévenin's theorem.
- Norton's theorem.
- Maximum power transfer theorem.
- Pi-T (Delta-Y) conversions.

5.0 Introduction

Superposition Theorem

Superposition allows the analysis of multi-source series-parallel circuits. Superposition can only be applied to networks that are linear and bilateral. Further, it cannot be used to find values for non-linear functions, such as power, directly (although power can be computed from the resulting voltage or current values). The basic idea is to determine the contribution of each source by itself, and then adding the results to get the final answer(s). Consider the circuit depicted in Figure 5A, below.

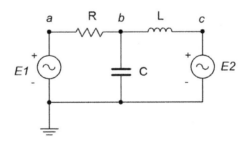

Figure 5A

Here we see two voltage sources, E1 and E2, driving a three element series-parallel network. As there are two sources, two derived circuits must be created; one using only E1 and the other using only E2. When considering a given source, all other sources are replaced by their ideal internal resistance. In the case of a voltage source, that's a short; and in the case of a current source, that's an open. When considering E1, E2 is replaced with a short. This leaves a fairly simple network where X_C and X_L are in parallel. This combination is in series with R and E1. Using basic series-parallel techniques, we can solve for desired quantities such as the current flowing through R or the voltage V_b. Be sure to indicate the current direction and voltage polarity with respect to the source being considered (here, that's left-to-right and positive). The process is then repeated for E2, shorting E1 and leaving us with R in parallel with X_C which is in turn in series with X_L and E2. Note that although in this version V_b is still positive, the current direction for R is now right-to-left. The numerical results from this version are added to those of the E1 version (minding polarities and directions) to achieve the final result. If power is needed, it can be computed from these currents and voltages. Note that superposition can work with a mix of current sources and voltage sources. The practical downside is that for large circuits using many sources, many derived circuits will need to be analyzed. For example, if there are three voltage sources and two current sources, then a total of five derived circuits will be created.

Source Conversions

For any simple voltage source consisting of an ideal voltage source with a series internal impedance, an equivalent current source may be created. Similarly, for any simple current source consisting of an ideal current source with a parallel internal impedance, an equivalent voltage source may be created. By "equivalent", we mean that both circuits will produce the same voltage and current to identical loads. Consider the simple voltage source of Figure 5B. It's equivalent current source is shown in Figure 5C.

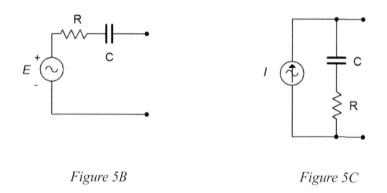

Figure 5B   *Figure 5C*

For reasons that will become apparent under the section on Thévenin's theorem below, the internal impedances of these two circuits must be identical if they are to behave identically. Knowing that, it is a straightforward process to find the required value of the other source. For example, given a voltage source, the maximum current that can be developed occurs when the load is shorted. This current is E/Z. Under that same load condition, all of the current from the current source version must be flowing through the load. Therefore, the value of the equivalent current source must be the maximum current of E/Z. Note that the resulting source normally will not have the same phase angle as the original source due to the phase angle of the associated impedance.

Similarly, if we start with a current source, an open load produces the maximum load voltage of $I \cdot Z$. Therefore, the equivalent voltage source must have a value of $I \cdot Z$. If a multi-source system is being converted (i.e., voltage sources in series or current sources in parallel), first combine the sources to arrive at the simplest source and then do the conversion. Do not convert the sources first and then combine them as this will produce series-parallel configurations rather than simple sources.

Judicious use of source conversions can sometimes simplify multi-source circuits by allowing converted sources to be combined, resulting in a single source. It is also possible to use superposition to find the resulting currents in a circuit that uses sources with different frequencies. In this instance, the equivalent circuits will have different reactance values. In fact, a single non-sinusoidal source can be analyzed using this method by treating the source as a series of superimposed sine waves with each sine source producing a new circuit with its own unique reactance values.

Dependent Sources

A *dependent source* is a current or voltage source whose value is not fixed (i.e., not independent) but rather which depends on some other circuit current or voltage. The general form for the value of a dependent source is *Y=kX* where *X* and *Y* are currents and/or voltages and *k* is the proportionality factor. For example, the value of a dependent voltage source may be a function of a current, so instead of the source being equal to, say, 10 volts, it could be equal to twenty times the current passing through a particular resistor, or *V=20I*.

There are four possible dependent sources: the voltage-controlled voltage source (VCVS), the voltage-controlled current source (VCCS), the current-controlled voltage source (CCVS), and the current-controlled current source (CCCS). The source and control parameters are the same for both the VCVS and the CCCS so *k* is unitless (although it may be given as volts/volt and amps/amp, respectively). For the VCCS and CCVS, *k* has units of amps/volt and volts/amp, respectively. These are referred to as the *transresistance* and *transconductance* of the sources with units of ohms and siemens.

The schematic symbols for dependent or controlled sources are usually drawn using a diamond instead of a circle. Also, there will be a secondary connection for the controlling current or voltage. Examples of a voltage-controlled sine voltage source (left) and a current-controlled current source (right) are shown in Figure 5D. On each of these symbols, the control element is shown to the left of the source. This portion is not always drawn on a schematic. Instead, the source simply may be labeled as a function, as in V = 0.02 I_X where I_X is the controlling current.

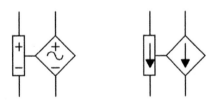

Figure 5D

Dependent sources are not "off-the-shelf" items in the same way that a battery or signal generator are. Rather, dependent sources are used to model the behavior of more complex devices. For example, a bipolar junction transistor commonly is modeled as a CCCS while a field effect transistor may be modeled as a VCCS. Similarly, many op amp circuits are modeled as VCVS systems. Solutions for circuits using dependent sources follow along the lines of those established for independent sources (i.e., the application of Ohm's law, KVL, KCL, etc.), however, the sources are now dependent on the remainder of the circuit which tends to complicate the analysis. In general, there are two possible configurations: isolated and coupled. An example of the isolated form is shown in Figure 5E.

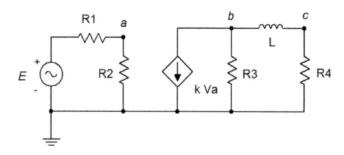

Figure 5E

In this example, the dependent source (center) does not interact with the subcircuit on the left driven by the independent source, thus it can be analyzed as two separate circuits. Solutions for this form are relatively straightforward in that the control value for the dependent source can be computed directly. The value is then substituted into the dependent source and the analysis continues as is typical. Sometimes it is convenient if the solution for a particular voltage or current is defined in terms of the control parameter rather than as a specific value (e.g., the voltage across a particular resistor might be 12 V_A instead of just 12 volts).

The second type of circuit is somewhat more complex in that the dependent source can affect the parameter that controls the dependent source. An example is shown in Figure 5F.

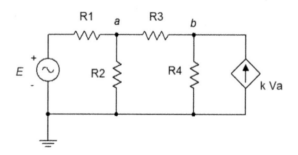

Figure 5F

In this example it should be obvious that the current from the dependent source can affect the voltage at node A, and it is this very voltage that in turn sets up the value of the current source. Circuits of this type can be analyzed using mesh or nodal analysis (nodal works well in this particular example) which are covered in the next section.

Finally, referring to the preceding section, it is possible to perform source conversions on dependent sources, within limits. The new source will remain a dependent source (e.g., VCVS to VCCS). This process is not applicable if the control parameter directly involves the internal impedance (i.e., is its voltage or current).

Thévenin's Theorem

Thévenin's theorem states that any two point linear network can be reduced to a simple voltage source (*Eth*) in series with an impedance *(Zth)* as shown in Figure 5G. This is a powerful analysis tool.

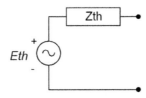

Figure 5G

The phrase "two point network" means that the circuit is cut in such a way that only two connections exist to the remainder of the circuit (i.e., one port). That remainder may be a single component or a large multi-component sub-circuit. As there are many ways to cut a typical circuit, there are many possible Thévenin equivalents. Consider the circuit shown in Figure 5H.

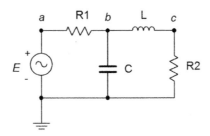

Figure 5H

Suppose we cut the circuit immediately to the left of R2. That is, we will find the Thévenin equivalent that drives R2. The first step is to make the cut, removing the remainder of the circuit (in this case, just R2). We then determine the open circuit output voltage. This is the maximum voltage that could appear between the cut points and is called the Thévenin voltage, E_{th}. This is shown in Figure 5I, following. In a circuit such as this, basic series-parallel analysis may be used to find E_{th} (note that due to the open, no current flows through L, thus no voltage is developed across it, and therefore E_{th} must equal the voltage developed across C).

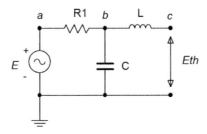

Figure 5I

The second part is finding the Thévenin impedance, Z_{th}. Beginning with the "cut" circuit, replace all sources with their ideal internal impedance (thus shorting voltage sources and opening current sources). From the perspective of the cut point, look back into the circuit and simplify to determine its equivalent resistance. This is shown in Figure 5J. Looking in from where the cut was made (right-to-left), we see that R1 and X_C are in parallel, and this combination is then in series with X_L. Thus, $Z_{th} = jX_L + (R1 \parallel -jX_C)$.

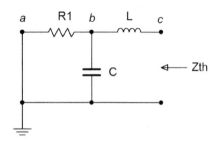

Figure 5J

As noted earlier, the original circuit could be cut in a number of different ways. We might, for example, want to determine the Thévenin equivalent that drives C in the circuit above. The cut appears below in Figure 5K.

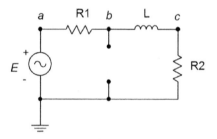

Figure 5K

Clearly, this will result in different values for both E_{th} and Z_{th}. For example, Z_{th} is now R1 $\parallel$ (R2 + jX_L).

Norton's Theorem

Norton's theorem is the current source version of Thévenin's theorem. In other words, a two point network can be reduced to a single current source with parallel internal impedance. The process is very similar. First, the Norton impedance is the same as the Thévenin impedance. Second, instead of finding the open circuit output voltage, the short circuit output current is found (again, the maximum value). This is the Norton current. If a Thévenin equivalent for a network can be created, then it must be possible to create a Norton equivalent. Indeed, if a Thévenin equivalent is found, a source conversion can be performed on it to yield the Norton equivalent.

Maximum Power Transfer Theorem

The maximum power transfer theorem states that in order to achieve the maximum power in a load, the load impedance must be equal to the complex conjugate of internal impedance of the source. The complex conjugate has the same real value but the opposite sign for the imaginary portion. This will cause the reactances to cancel thus producing maximal current. No other value of load impedance will produce a higher load power. For the circuit of Figure 5L, this means that Z_{load} (i.e., $R_{load} - jX_{Cload}$) must equal the complex conjugate of Z_{source} (i.e., $R_i + jX_{Li}$). In other words, R_{load} must equal R_i and $|X_{Cload}|$ must equal $|X_{Li}|$. A similar situation exists in Figure 5M where the source is capacitive, thus requiring an inductive load.

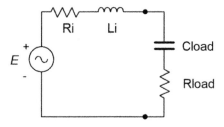

Figure 5L, Inductive Source

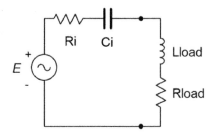

Figure 5M, Capacitive Source

While this produces the maximum load power, it does not produce maximum load current or maximum load voltage. In fact, this condition produces a load voltage and current that are half of their maximums. Their product, however, is at the maximum. Further, efficiency at maximum load power is only 50%. Note that values of R_{load} greater than R_i will achieve higher efficiency but at reduced load power.

As any linear two point network can be reduced to something like Figure 5L or 5M by using Thévenin's theorem, combining the two theorems allows the maximum power conditions for any impedance in a complex circuit to be determined.

Pi-T (Delta-Y) Conversions

Certain component configurations, such as bridged networks, cannot be reduced to a simple impedance using basic series-parallel conversion techniques. One method for simplification involves converting sections into more convenient forms. The configurations in question are three point networks containing three resistors. Due to the manner in which they drawn, they are referred to as pi (π) networks and T networks. Alternately, if they are slightly redrawn they are known as delta (Δ) networks and Y networks. These networks are shown in Figure 5N.

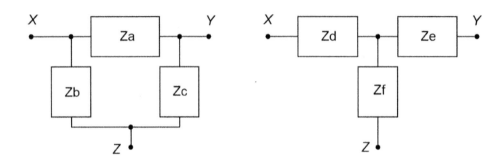

Figure 5N, Pi-T (π-T)

It is possible to convert back and forth between delta and Y networks. That is, for every delta network, there exists a Y network such that the impedances seen between the X, Y and Z terminals are identical, and vice versa. Consequently, one configuration can replace another in order to simplify a larger circuit.

To convert from delta to Y:

$Zd = (Za \cdot Zb)/(Za+Zb+Zc)$
$Ze = (Za \cdot Zc)/(Za+Zb+Zc)$
$Zf = (Zb \cdot Zc)/(Za+Zb+Zc)$

To convert from Y to delta:

$Za = (Zd \cdot Ze + Ze \cdot Zf + Zd \cdot Zf)/(Zf)$
$Zb = (Zd \cdot Ze + Ze \cdot Zf + Zd \cdot Zf)/(Ze)$
$Zc = (Zd \cdot Ze + Ze \cdot Zf + Zd \cdot Zf)/(Zd)$

5.1 Exercises

Analysis

1. For the circuit shown in Figure 5.1, use Superposition to find v_b.

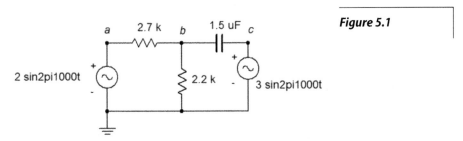

Figure 5.1

2. For the circuit shown in Figure 5.1, use Superposition to find the current through the capacitor.

3. Use Superposition to find the current through the 82 Ω resistor. For the circuit shown in Figure 5.2.

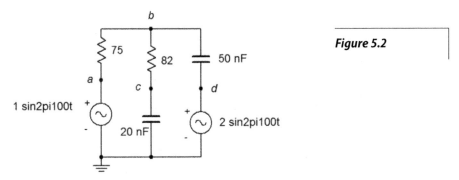

Figure 5.2

4. Use Superposition to find v_b and v_{cd} for the circuit shown in Figure 5.2.

5. In the circuit of Figure 5.3, use Superposition to find v_b. Source one is one volt peak and source two is two volts peak.

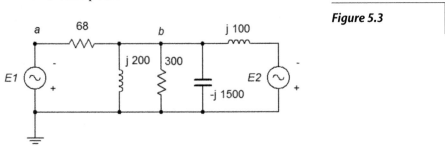

Figure 5.3

6. In the circuit of Figure 5.3, use Superposition to find the currents through the two inductors. Source one is two volts peak and source two is three volts peak.

7. Use Superposition to find the current through the 2.2 kΩ resistor for the circuit of Figure 5.4. $E1 = 1\angle 0°$ and $E2 = 10\angle 90°$.

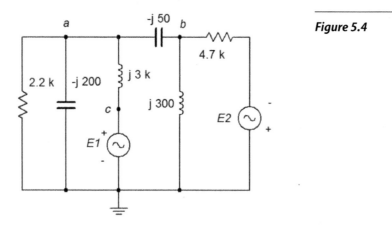

Figure 5.4

8. Use Superposition to find v_{ab} for In the circuit of Figure 5.4. $E1 = 1\angle 0°$ and $E2 = 2\angle 45°$.

9. In the circuit of Figure 5.5, use Superposition to find v_b and v_{cd}. The sources are in phase.

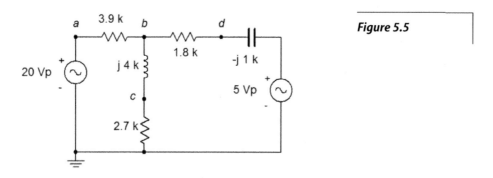

Figure 5.5

10. In the circuit of Figure 5.5, use Superposition to find the current through the capacitor. The sources are in phase.

11. Use Superposition to find the two source currents for In the circuit of Figure 5.6. Source one is 100 mV peak and source two is 500 mV peak (in phase).

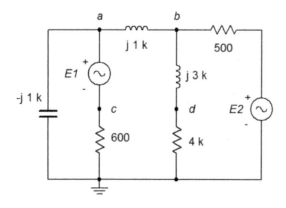

Figure 5.6

66

12. Use Superposition to find v_{cd} for the circuit of Figure 5.6. Source one is 100 mV peak and source two is 1 V peak (in phase).

13. In the circuit of Figure 5.7, use Superposition to find v_{ab}.

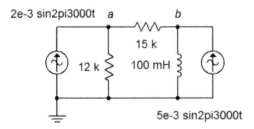

Figure 5.7

14. In the circuit of Figure 5.7, use Superposition to find the current through the 15 kΩ resistor.

15. In the circuit of Figure 5.8, use Superposition to find v_{ab}.

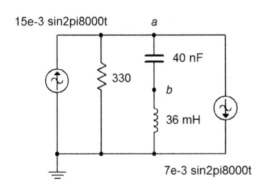

Figure 5.8

16. In the circuit of Figure 5.8, use Superposition to find the current flowing through the resistor.

17. For the circuit of Figure 5.9, use Superposition to find v_a and v_b. The sources are in phase.

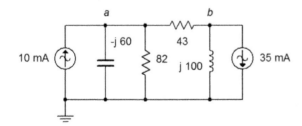

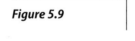

Figure 5.9

18. For the circuit of Figure 5.9, use Superposition to find the currents through the inductor and capacitor. The sources are in phase.

19. Use Superposition in the circuit of Figure 5.10 to find the currents through the inductor and capacitor. $I1 = 1\angle 45°$ and $I2 = 2\angle 45°$.

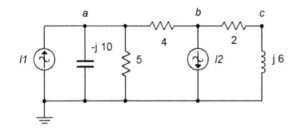

Figure 5.10

20. Use Superposition in the circuit of Figure 5.10 to find v_{ab} and v_{bc}. $I1 = 1\angle 0°$ and $I2 = 2\angle 90°$.

21. In the circuit of Figure 5.11, Use Superposition to find v_{bc}. $I1 = 10\angle 0°$ and $I2 = 6\angle 0°$.

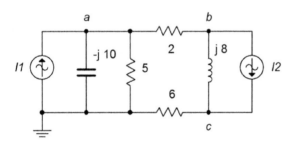

Figure 5.11

22. In the circuit of Figure 5.11, Use Superposition to find the current flowing through the 2 Ω resistor. $I1 = 4\angle 120°$ and $I2 = 6\angle 0°$.

23. Use Superposition to determine the current of source E in the circuit of Figure 5.12. $E = 40\angle 180°$ and $I = 20\text{E}{-}3\angle 0°$.

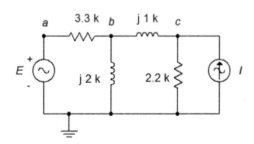

Figure 5.12

24. Use Superposition to determine v_{ac} in the circuit of Figure 5.12. $E = 28\angle 0°$ and $I = 8\text{E}{-}3\angle{-}180°$.

25. Use Superposition to determine v_b in the circuit of Figure 5.13. $I = 3E{-}3\angle 0°$ and $E = 9\angle 0°$.

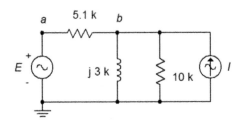

Figure 5.13

26. Use Superposition to determine the inductor current in the circuit of Figure 5.13. $I = 4E{-}3\angle 0°$ and $E = 18\angle{-}45°$.

27. For the circuit of Figure 5.14, use Superposition to determine the inductor current. $I = 100E{-}3\angle 0°$ and $E = 26\angle 0°$.

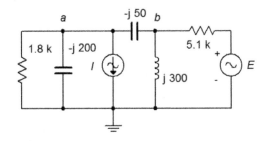

Figure 5.14

28. For the circuit of Figure 5.14, use Superposition to determine v_{ab}. $I = 50E{-}3\angle 0°$ and $E = 18\angle 90°$.

29. Use Superposition to determine v_{ab} in the circuit of Figure 5.15. $I = 10E{-}3\angle 0°$ and $E = 12\angle 0°$.

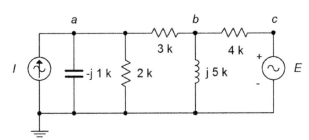

Figure 5.15

30. Use Superposition to determine the capacitor current in the circuit of Figure 5.15. $I = 5E{-}3\angle 0°$ and $E = 18\angle 120°$.

69

31. Determine v_b in the circuit of Figure 5.16 if the source $I1 = 20\text{E}{-}3\angle 0°$.

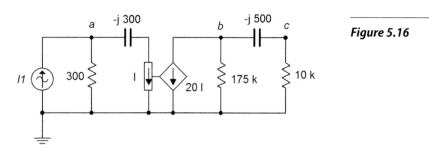

Figure 5.16

32. Determine the current through the 10 kΩ resistor in the circuit of Figure 5.16 if $I1 = 10\text{E}{-}3\angle{-}90°$.

33. Determine the current through the 5 kΩ resistor in the circuit of Figure 5.17 if $E = 10\angle 0°$.

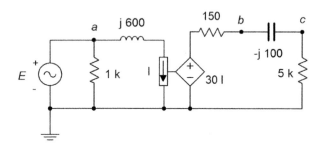

Figure 5.17

34. Determine v_c in the circuit of Figure 5.17 if the source $E = 3\angle 120°$.

35. In the circuit of Figure 5.18, determine v_c if the source $E = 8\angle 90°$.

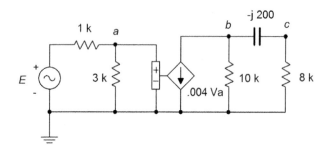

Figure 5.18

36. In the circuit of Figure 5.18, determine the capacitor current if the source $E = 12\angle 0°$.

37. In the circuit of Figure 5.19, determine the current flowing into the 1 k resistor if the source $E = 6\angle 0°$.

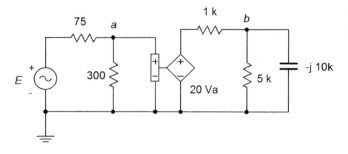

Figure 5.19

38. In the circuit of Figure 5.19, determine v_b if the source $E = 12\angle-90°$.

39. Determine v_a and v_b in the circuit of Figure 5.20 if the source $I = 2E{-}3\angle 0°$.

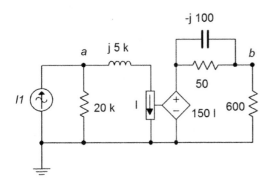

Figure 5.20

40. In the circuit of Figure 5.20, determine the current flowing into the 600 Ω resistor if $I = 1E{-}3\angle 180°$.

41. For the circuit of Figure 5.21, determine the Thévenin equivalent that drives the 20 nF capacitor.

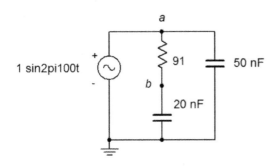

Figure 5.21

42. Given the circuit of Figure 5.21, determine the Norton equivalent that drives the 20 nF capacitor.

43. For the circuit of Figure 5.22, determine the Thévenin and Norton equivalents that drive the 600 Ω resistor if the source $E = 12\angle 0°$.

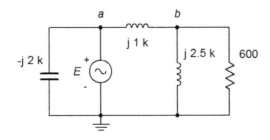

Figure 5.22

44. Given the circuit of Figure 5.22, determine the Thévenin equivalent that drives the $j1$ kΩ inductive reactance if $E = 9\angle 0°$.

45. Given the circuit of Figure 5.22, determine the Norton equivalent that drives the $j2.5$ kΩ inductive reactance if $E = 24\angle 45°$.

46. Use Thévenin's Theorem to find v_b in the circuit of Figure 5.22 if $E = 18\angle 0°$.

47. Use Thévenin's Theorem to find v_b in the circuit of Figure 5.23.

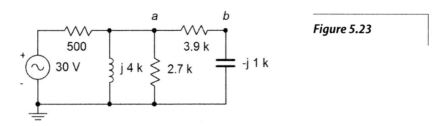

Figure 5.23

48. Determine the Thévenin equivalent that drives the 3.9 kΩ + j1 kΩ combo in the circuit of Figure 5.23. Does this combo's impedance achieve maximum load power? If not, what combo will achieve maximum power and what is the resulting power?

49. Determine the Norton equivalent that drives the 500 Ω resistor in the circuit of Figure 5.23. Determine the value of component(s) that when placed in series with the 500 Ω resistor will achieve maximum load power (i.e., for the combo as the load).

50. For the circuit of Figure 5.24, determine the Thévenin and Norton equivalents that drive the combo of 36 Ω + j100 Ω. Does this combo achieve maximum load power? If not, what combo will achieve maximum power and what is the resulting power?

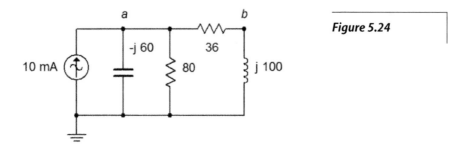

Figure 5.24

51. For the circuit of Figure 5.25, determine the Thévenin and Norton equivalents that drive the combo of 300 Ω in parallel with $-j$1500 Ω. Does this combo achieve maximum load power? If not, what combo will achieve maximum power and what is the resulting power? $E = 120\angle 0°$.

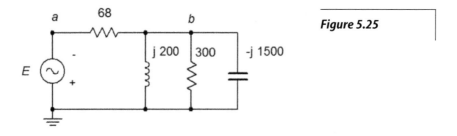

Figure 5.25

52. For the circuit of Figure 5.26, determine the Thévenin and Norton equivalents that drive the combo of 4.7 kΩ in parallel with $j300$ Ω. Does this combo achieve maximum load power? If not, what combo will achieve maximum power and what is the resulting power? $I = 200\text{E}{-}3\angle 0°$.

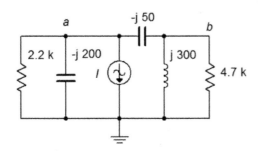

Figure 5.26

53. Determine the equivalent Y (T) network for the circuit of Figure 5.27. R1 = R2 = R3 = 10 kΩ and $X_{L1} = X_{L2} = X_{L3} = j10$ kΩ.

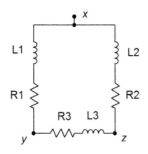

Figure 5.27

54. Determine the equivalent Y (T) network for the circuit of Figure 5.28.

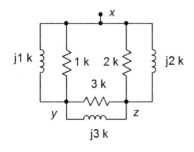

Figure 5.28

55. Determine the equivalent Y (T) network for the circuit of Figure 5.29. R1 = R2 = R3 = 4 kΩ and $X_{C1} = X_{C2} = X_{C3} = -j3$ kΩ.

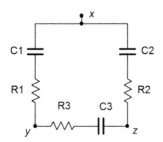

Figure 5.29

73

56. Determine the equivalent Y (T) network for the circuit of Figure 5.30.

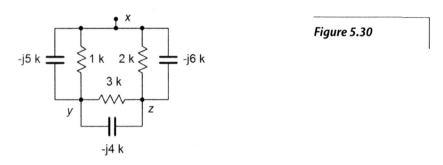

Figure 5.30

57. Determine the equivalent delta (pi) network for the circuit of Figure 5.31.

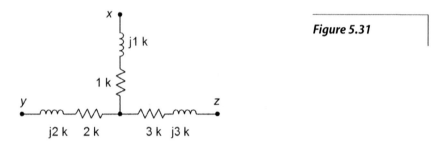

Figure 5.31

58. Determine the equivalent delta (pi) network for the circuit of Figure 5.32.

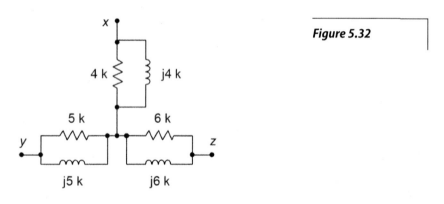

Figure 5.32

59. Determine the equivalent delta (pi) network for the circuit of Figure 5.33.

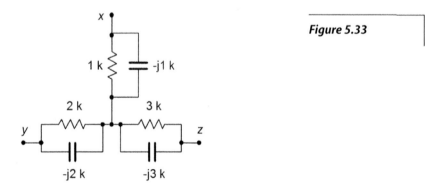

Figure 5.33

60. Determine the equivalent delta (pi) network for the circuit of Figure 5.34.

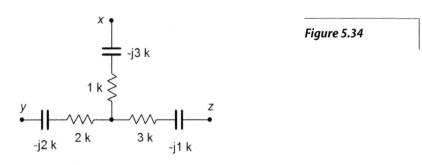

Figure 5.34

61. Find voltage v_{bc} in the circuit of Figure 5.35 through the use of one or more delta-Y conversions. $E = 10\angle 0°$, R1 = 1 kΩ, R2 = 2 kΩ, R3 = 3 kΩ, $X_C = -j4$ kΩ and $X_L = j8$ kΩ.

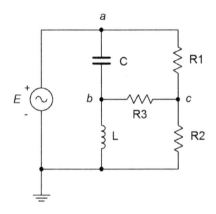

Figure 5.35

62. Find voltage v_{bc} in the circuit of Figure 5.36 through the use of one or more delta-Y conversions. $E = 20\angle 0°$, R1 = 1 kΩ, R2 = 8 kΩ, R3 = 3 kΩ, $X_C = -j4$ kΩ and $X_L = j2$ kΩ.

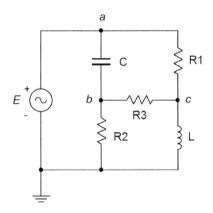

Figure 5.36

Design

63. Design an equivalent current source for Figure 5.37. $E = 12\angle 90°$, $R = 1$ kΩ and $X_C = -j200$ Ω. The source frequency is 10 kHz.

Figure 5.37

64. Design an equivalent current source for Figure 5.37 if $E = 10\angle 0°$, $R = 2.2$ kΩ and $C = 100$ nF. The source frequency is 1 kHz.

65. Design an equivalent current source for Figure 5.38 if $E = 1\angle 0°$, $R = 600$ Ω and $L = 2$ mH. The source frequency is 20 kHz.

Figure 5.38

66. Design an equivalent current source for Figure 5.38. $E = 2\angle 90°$, $R = 10$ kΩ and $X_L = j900$ Ω.

67. Design an equivalent voltage source for Figure 5.39. $I = 300\text{E}{-3}\angle 0°$, $R = 4.3$ kΩ and $X_C = -j5$ kΩ.

Figure 5.39

68. Design an equivalent voltage source for Figure 5.39 if $I = 100\text{E}{-3}\angle 120°$, $R = 75$ Ω and $L = 1$ mH. The source frequency is 10 kHz.

69. Design an equivalent voltage source for Figure 5.40 if $I = 10E-3\angle 0°$, $R = 9.1$ kΩ and $L = 5$ mH. The source frequency is 100 kHz.

Figure 5.40

70. Design an equivalent voltage source for Figure 5.40 if $I = 50E-3\angle 0°$, $R = 560$ Ω and $X_L = j350$ Ω.

71. Reconfigure the circuit of Figure 5.41 so that it uses only voltage sources. Express all new component and source values in terms of the original labels.

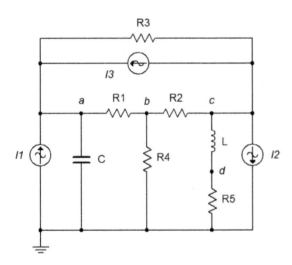

Figure 5.41

72. Reconfigure the circuit of Figure 5.42 so that it uses only current sources. Express all new component and source values in terms of the original labels.

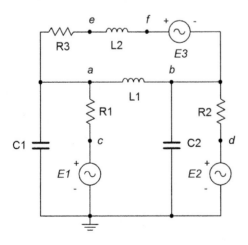

Figure 5.42

73. Redesign the circuit of Figure 5.1 so that it uses only current sources and produces the same node voltages as the original circuit.

74. Consider the 600 Ω resistor to be the load in Figure 5.22. Determine a new value for the load in order to achieve maximum load power. Also determine the maximum load power.

75. Using Thévenin's Theorem with the circuit of Figure 5.23, determine a new value of capacitive reactance such that it cancels the Thévenin reactance.

76. Using Norton's Theorem with the circuit of Figure 5.24, determine a new value of inductive reactance such that the inductor current is 1 mA in magnitude.

Challenge

77. Redesign the circuit of problem 7 (Figure 5.4) so that it uses only current sources and produces the same node voltages as the original circuit.

78. In the circuit of Figure 5.43, use any method or combination of methods to determine v_{ab}. $I1 = 0.05\angle 0°$, $I2 = 0.1\angle 0°$ and $I3 = 0.2\angle 90°$.

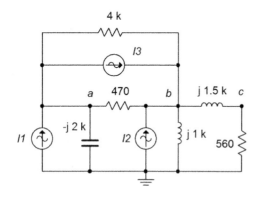

Figure 5.43

79. In the circuit of Figure 5.44, use any method or combination of methods to determine v_{ab}. $E1 = 5\angle 0°$, $E2 = 10\angle 90°$ and $E3 = 15\angle 0°$.

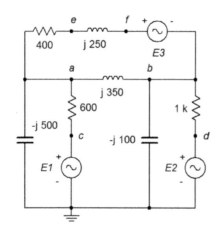

Figure 5.44

80. Determine v_{ab} in the circuit of Figure 5.45. $E = 10\angle 0°$

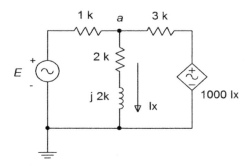

Figure 5.45

81. Use any method or combination of methods in the circuit of Figure 5.46 to determine v_{ad}. $E1 = 90\angle 0°$, $E2 = 120\angle 0°$ and $I = 400E{-}3\angle 180°$.

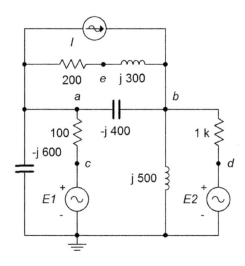

Figure 5.46

82. Consider the 2.7 kΩ + $j4$ kΩ combo to be the load in Figure 5.5. Determine if this value achieves maximum load power. If not, determine a new value for the load in order to achieve maximum load power. Also determine the maximum load power.

83. In the circuit of Figure 5.47, assume the source E is 120 volts RMS at 60 Hz. Determine the value for the load, Z, that will produce maximum load power. Express Z in terms of a resistor and either an inductor or capacitor. Further, specify both the series and parallel equivalents for the load.

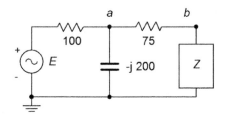

Figure 5.47

84. Convert the circuit of Figure 5.48 into the equivalent delta configuration.

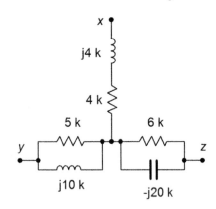

Figure 5.48

85. Find the Thévenin equivalent looking into nodes a and b for the circuit of Figure 5.49. $I = 4\angle 0°$.

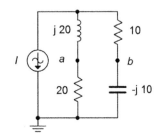

Figure 5.49

86. Find voltage v_{bc} in the circuit of Figure 5.50 through the use of one or more delta-Y conversions.
$E = 100\angle 0°$, $R1 = R2 = 2\ k\Omega$, $R3 = 3\ k\Omega$, $R4 = 10\ k\Omega$, $R5 = 5\ k\Omega$, $X_{C1} = X_{C2} = -j2\ k\Omega$.

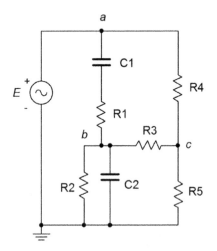

Figure 5.50

87. Given the circuit of Figure 5.51, determine an equivalent circuit using a single voltage source.
$E1 = 100\angle 0°$, $E2 = 60\angle 180°$, $E3 = 40\angle 90°$, $E4 = 75\angle 0°$.

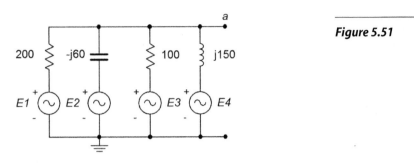

Figure 5.51

Simulation

88. Verify the voltage computed for problem 1 by running a transient analysis.

89. Verify the voltage computed for problem 4 by running a transient analysis.

90. Using multiple transient analysis simulations, compare the original circuit of problem 64 to its converted equivalent. Do this by connecting various components to the output terminals, trying several different impedance values and checking to see if the two circuits always produce the same voltage across this impedance.

91. Using multiple transient analysis simulations, compare the original circuit of problem 69 to its converted equivalent. Do this by connecting various components to the output terminals, trying several different impedance values and checking to see if the two circuits always produce the same voltage across this impedance.

92. Run a transient analysis to verify the design of problem 75.

93. Run a transient analysis to verify the design of problem 76.

6 Mesh and Nodal Analysis

This section covers:
- Series-parallel RLC circuits using multiple voltage and/or current sources via mesh and nodal analyses.

6.0 Introduction

Mesh Analysis

Mesh analysis uses Kirchhoff's voltage law (KVL) to create a series of loop equations that can be solved for mesh currents. The current through any particular component may be a mesh current or a combination of mesh currents. Circuits using complex series-parallel arrangements with multiple voltage and/or current sources may solved using this technique.

Consider the circuit of Figure 6A. We begin by designating a series of current loops. These loops should be minimal in size and together cover all components at least once. By convention, the loops are drawn clockwise. There is nothing magical about them being clockwise, it is just a matter of consistency. In the circuit of Figure 6A we have two loop currents, I1 and I2. Note that all components exist in at least one loop (and sometimes in more than one loop, like C). Depending on circuit values, one or more of these loop directions may in fact be opposite of reality. This is not a problem. If this is the case, the currents will show up as negative values, and thus we know that they're really flowing counter-clockwise.

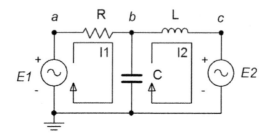

Figure 6A

We begin by writing KVL equations for each loop.

Loop 1: $E1 = V$ across $R + V$ across X_C
Loop 2: $-E2 = V$ across $X_C + V$ across X_L (*E2* is negative as *I2* is drawn flowing out of its negative terminal)

Expand the voltage terms using Ohm's law.

Loop 1: $E1 = I1 \cdot R + (I1 - I2)(-jX_C)$
Loop 2: $-E2 = I2 \cdot (jX_L) + (I2 - I1)(-jX_C)$

Multiplying out and collecting terms yields:

Loop 1: $E1 = (R - jX_C) I1 - (-jX_C) \cdot I2$
Loop 2: $-E2 = -(-jX_C) \cdot I1 + (jX_L - jX_C) I2$

As the component values and source voltages are known, we have two equations with two unknowns. These can be solved for $I1$ and $I2$ using simultaneous equation solution techniques such as determinants or Gauss-Jordan elimination. These equations also can be obtained through inspection. Simply focus on one loop and ask the following questions: what is the total source voltage in this loop? This yields the voltage constant. Then simply sum the resistance and reactance values in the loop under inspection. This yields the coefficient for that current term. For the other current coefficients, sum the resistances and reactances that are in common between the loop under inspection and the other loops (e.g., for loop 1, X_C is in common with loop 2). These values will always be negative. As a crosscheck, the set of equations produced must exhibit *diagonal symmetry*, that is, if a diagonal is drawn from the upper left to the lower right through the *IZ* pairs, then the coefficients found above the diagonal will have to match those found below the diagonal. In the example above, note the matching "$(-jX_C)$" coefficients in the final pair of equations.

While it is possible to extend this technique to include current sources, it may be easier and less error-prone to convert the current sources into voltage sources and continue with the direct inspection method outline above. In closing, it is important to remember that the number of loops determines the number of equations to be solved.

Nodal Analysis

Nodal analysis uses KCL to create a series of node equations that can be solved for node voltages. In some respects it is similar to mesh analysis. We will examine two variations; one using voltage sources, the other using current sources.

Consider the circuit shown in Figure 6B, following. We begin by labeling connection points and assigning current directions. These directions are chosen arbitrarily and may be opposite of reality. If so, their values will ultimately show up as negative.

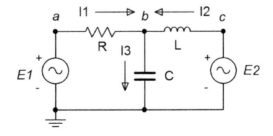

Figure 6B

Write a current summation equation for each summing node, except for ground. In this circuit there is only one node where currents combine, node b.

$I1 + I2 = I3$

Describe these currents in terms of the node voltages and components. For example, $I3$ is the node b voltage divided by $-jX_C$ while $I1$ is the voltage across R divided by R. This voltage is $V_a - V_b$.

$(V_a - V_b)/R + (V_c - V_b)/jX_L = V_b/(-jX_C)$

Noting that $V_a = E1$ and $V_c = E2$, with a little algebra this can be reduced to:

$E1(1/R) + E2(1/jX_L) = V_b(1/R + 1/jX_L + 1/(-jX_C))$

All quantities are known except for V_b. If there had been more nodes, there would have been an equal number of equations.

For current sources, a more direct approach is possible. Consider the circuit of Figure 6C. We start as before, identifying nodes and labeling currents. We then write the current summation equations at each node (except for ground). We consider currents entering a node as positive and exiting as negative.

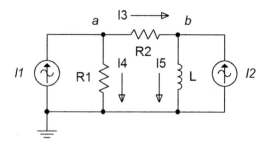

Figure 6C

Node a: $I1 = I3 + I4$
Node b: $I3 + I2 = I5$, and rearranging in terms of the fixed source,
Node b: $I2 = -I3 + I5$

The currents are then described by their Ohm's law equivalents:

Node a: $I1 = (V_a - V_b)/R2 + V_a/R1$
Node b: $I2 = -(V_a - V_b)/R2 + V_b/jX_L$

Expanding and collecting terms yields:

Node a: $I1 = (1/R1 + 1/R2)V_a - (1/R2)V_b$
Node b: $I2 = -(1/R2)V_a + (1/R2 + 1/jX_L)V_b$

As the component values and currents are known, the node voltages may be solved for using simultaneous equation solution techniques. There will be as many equations as node voltages.

Like mesh analysis, there is a method to generate the equations by inspection. For the node under inspection, sum all of the current sources connected to it to obtain the current constant. The conductance term for that node will be the sum of all of the conductances connected to that node. For the other node conductances, determine the conductances between the node under inspection and these other nodes. These terms will all be negative and once again, the set of equations thus produced must exhibit diagonal symmetry (note the "−1/R2" coefficients for the final set of equations, above). For example, focusing on node a above, we find the fixed current source $I1$ feeding it (entering, therefore positive). The conductances directly connected to node a are 1/R1 and 1/R2, yielding the coefficient for V_a. The only conductance common between nodes a and b is 1/R2, yielding the V_b coefficient.

Using Source Conversions

Given circuits with voltage sources, it may be easier to convert them to current sources and then apply the inspection technique rather than using the general approach outlined initially. There is one trap to watch out for when using source conversions, and this also applies to mesh analysis: **the voltage across or current through a converted component will most likely not be the same as the voltage or current in the original circuit.** This is because the location of the converted component will have changed. For example, the circuit of Figure 6C could be solved using mesh analysis by converting the current sources and their associated resistances or reactances into current sources. That is, $I1$/R1 would be converted into a source $E1$ with a series resistor R1. Although R1 still connects to node a, the other end no longer connects to ground. Rather, it connects to the new $E1$. Therefore, the voltage drop across R1 in the converted circuit is not likely to equal the voltage drop seen across R1 in the original circuit (the only way they would be equal is if $E1$ turned out to be 0). In the converted circuit, node a has not changed from the original, so the original voltage across R1 can be determined via V_a.

Dependent Sources

If the circuit includes dependent sources, either the general form of nodal or mesh may be used for analysis. The dependent source(s) will contribute terms that include the controlling parameter(s) so some additional effort will be in order. To illustrate, consider the circuit of Figure 6D. We shall use nodal analysis.

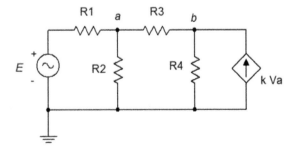

Figure 6D

We begin by defining current directions. Assume that the currents through R1 and R3 are flowing into node A, the current through R2 is flowing out of node A, and the current through R4 is flowing out of node B. We shall number the branch currents to reflect the associated resistor. The resulting KCL equations are:

Node a: $I1 + I3 = I2$
Node b: $k\, V_a = I3 + I4$

The currents are then described by their Ohm's law equivalents:

Node a: $(E - V_a)/R1 + (V_b - V_a)/R3 = V_a/R2$
Node b: $k\, V_a = (V_b - V_a)/R2 + V_b/R4$

Expanding terms yields:

Node a: $E/R1 - V_a/R1 + V_b/R3 - V_a/R3 = V_a/R2$
Node b: $k\, V_a = V_b/R2 - V_a/R2 + V_b/R4$

Collecting terms and simplifying yields:

Node a: $E/R1 = (1/R1 + 1/R2 + 1/R3)\, V_a - (1/R3)\, V_b$
Node b: $0 = - (k + 1/R2)\, V_a + (1/R2 + 1/R4)\, V_b$

Values for the resistors, k and E are known, so the analysis proceeds as usual.

6.1 Exercises

Analysis

1. Given the circuit in Figure 6.1, write the mesh loop equations and determine v_b.

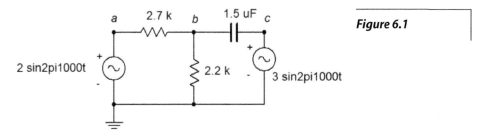

Figure 6.1

2. Use mesh analysis to find the current through the 2.7 kΩ resistor in the circuit of Figure 6.1.
3. Use mesh analysis to find the current through the 75 Ω resistor in the circuit of Figure 6.2.

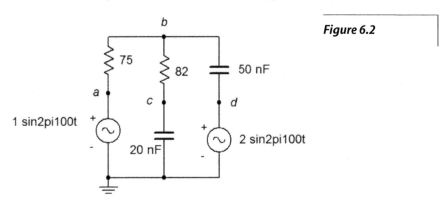

Figure 6.2

4. Given the circuit in Figure 6.2, write the mesh loop equations and determine v_c.
5. Given the circuit in Figure 6.3, write the mesh loop equations and determine v_b.

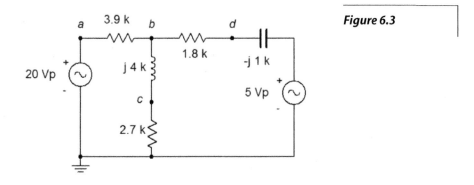

Figure 6.3

87

6. Use mesh analysis to find the current through the 1.8 kΩ resistor in the circuit of Figure 6.3.

7. Use mesh analysis to find the current through the $j200\ \Omega$ inductor in Figure 6.4. $E1 = 1\angle 0°$, $E2 = 2\angle 0°$.

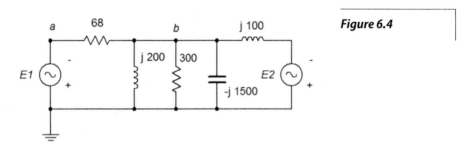

Figure 6.4

8. Given the circuit in Figure 6.4, write the mesh loop equations and determine v_b. Consider using parallel simplification first. $E1 = 36\angle -90°$, $E2 = 24\angle -90°$.

9. Given the circuit in Figure 6.5, use mesh analysis to determine v_{cd}. $E1 = 0.1\angle 0°$, $E2 = 0.5\angle 0°$.

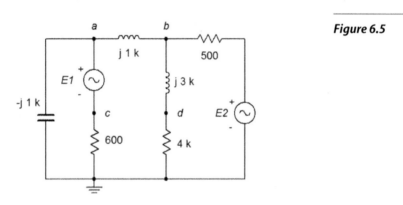

Figure 6.5

10. Use mesh analysis to find the current through the 600 Ω resistor in the circuit of Figure 6.5. $E1 = 9\angle 0°$, $E2 = 5\angle 40°$.

11. Use mesh analysis to find the current through the $-j200\ \Omega$ capacitor in the circuit of Figure 6.6. $E1 = 18\angle 0°$, $E2 = 12\angle 90°$.

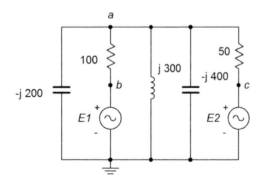

Figure 6.6

12. Given the circuit in Figure 6.6, use mesh analysis to determine v_{ac}. $E1 = 1\angle 0°$, $E2 = 500\text{E}{-}3\angle 0°$.

13. Given the circuit in Figure 6.7, use mesh analysis to determine v_c. $E1 = 10\angle{-}180°$, $E2 = 25\angle 0°$.

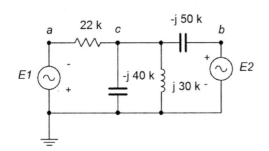

Figure 6.7

14. Use mesh analysis to find the current through the 22 kΩ resistor in the circuit of Figure 6.7. $E1 = 24\angle 0°$, $E2 = 36\angle 0°$.

15. Use mesh analysis to find the current through the $j300$ Ω inductor in Figure 6.8. $E1 = 1\angle 0°$, $E2 = 10\angle 90°$.

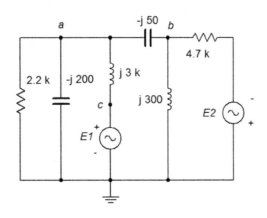

Figure 6.8

16. Given the circuit in Figure 6.8, use mesh analysis to determine v_a. $E1 = 100\angle 0°$, $E2 = 90\angle 0°$.

17. Given the circuit in Figure 6.9, use mesh analysis to determine v_{bc}. $E = 10\angle 0°$, $R1 = 1$ kΩ, $R2 = 2$ kΩ, $R3 = 3$ kΩ, $X_C = -j4$ kΩ, $X_L = j8$ kΩ.

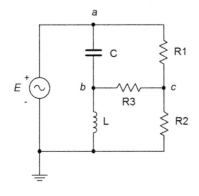

Figure 6.9

18. Use mesh analysis to find the current through resistor R3 in the circuit of Figure 6.9. $E = 20\angle 0°$, R1 = 10 kΩ, R2 = 30 kΩ, R3 = 1 kΩ, $X_C = -j15$ kΩ, $X_L = j20$ kΩ.

19. Use mesh analysis to find the current through resistor R3 in Figure 6.10. $E = 60\angle 0°$, R1 = 1 kΩ, R2 = 2 kΩ, R3 = 3 kΩ, $X_C = -j10$ kΩ, $X_L = j20$ kΩ.

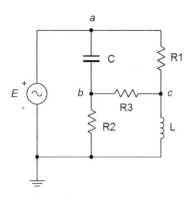

Figure 6.10

20. Given the circuit in Figure 6.10, use mesh analysis to determine v_{bc}. $E = 120\angle 90°$, R1 = 100 kΩ, R2 = 20 kΩ, R3 = 10 kΩ, $X_C = -j5$ kΩ, $X_L = j20$ kΩ.

21. Given the circuit in Figure 6.11, use mesh analysis to determine v_b. Consider using source conversion. $E = 12\angle 0°$, $I = 10E{-}3\angle 0°$.

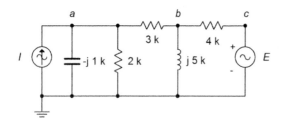

Figure 6.11

22. Use mesh analysis to find the current through the 3 Ω resistor in the circuit in Figure 6.11. Consider using source conversion. $E = 15\angle 90°$, $I = 10E{-}3\angle 0°$.

23. Use mesh analysis to find the current through the 2.2 kΩ resistor in the circuit in Figure 6.12. $E = 3.3\angle 0°$, $I = 2.1E{-}3\angle 0°$.

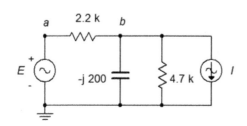

Figure 6.12

24. Given the circuit in Figure 6.12, use mesh analysis to determine v_b. $E = 10\angle 0°$, $I = 30E{-}3\angle 90°$.

25. Given the circuit in Figure 6.13, use nodal analysis to determine v_{ab}.

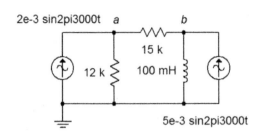

Figure 6.13

26. Use nodal analysis to find the current through the 100 mH inductor in the circuit of Figure 6.13.

27. Use nodal analysis to find the current through the 330 Ω resistor in the circuit of Figure 6.14.

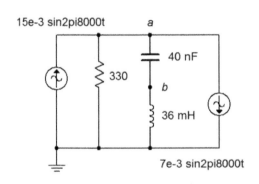

Figure 6.14

28. Given the circuit in Figure 6.14, write the node equations and determine v_b.

29. Given the circuit in Figure 6.15, use nodal analysis to determine v_c. $I1 = 3\angle 0°$, $I2 = 900E{-}3\angle 0°$

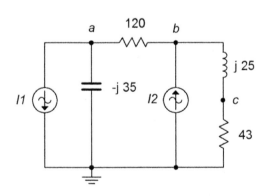

Figure 6.15

30. Use nodal analysis to find the current through the 120 Ω resistor in the circuit of Figure 6.15. $I1 = 500E{-}3\angle 90°$, $I2 = 1.6\angle 0°$.

31. Use nodal analysis to find the current through the 43 Ω resistor in the circuit of Figure 6.16. The sources are in phase.

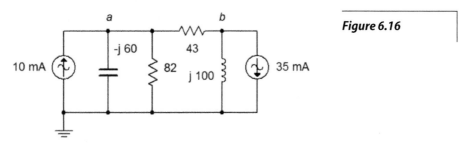

Figure 6.16

32. Given the circuit in Figure 6.16, use nodal analysis to determine v_b. The sources are in phase.

33. Given the circuit in Figure 6.17, determine v_c. $I1 = 3\angle 0°$, $I2 = 2\angle 0°$.

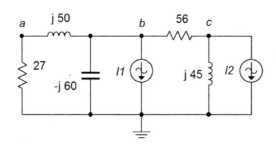

Figure 6.17

34. Use nodal analysis to find the current through the $j45$ Ω inductor in the circuit of Figure 6.17. $I1 = 2\angle 0°$, $I2 = 1.5\angle 60°$.

35. Use nodal analysis to find the current through the 4 Ω resistor in the circuit of Figure 6.18. $I1 = 1\angle 45°$, $I2 = 2\angle 45°$.

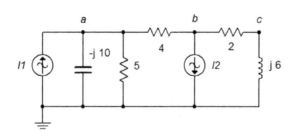

Figure 6.18

36. Given the circuit in Figure 6.18, use nodal analysis to determine v_c. $I1 = 6\angle 30°$, $I2 = 4\angle 0°$.

37. Given the circuit in Figure 6.19, use nodal analysis to determine v_{ac}. $I1 = 10\angle 0°$, $I2 = 6\angle 0°$.

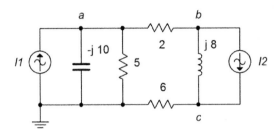

Figure 6.19

38. Use nodal analysis to find the current through the $j8\ \Omega$ inductor in the circuit of Figure 6.19. $I1 = 3\angle 0°$, $I2 = 5\angle 30°$.

39. Use nodal analysis to find the current through the 22 Ω resistor in the circuit of Figure 6.20. $I1 = 800\text{E}{-}3\angle 0°$, $I2 = 2.5\angle 0°$, $I3 = 2\angle 20°$.

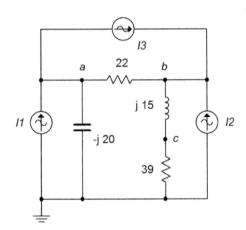

Figure 6.20

40. Given the circuit in Figure 6.20, use nodal analysis to determine v_c. $I1 = 4\angle 90°$, $I2 = 10\angle 120°$, $I3 = 5\angle 0°$.

41. Given the circuit in Figure 6.21, use nodal analysis to determine v_c. $I1 = 3\text{E}{-}3\angle 0°$, $I2 = 10\text{E}{-}3\angle 0°$, $I3 = 2\text{E}{-}3\angle 0°$.

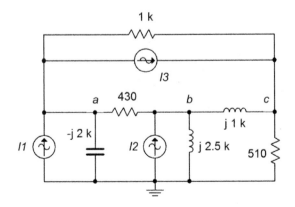

Figure 6.21

93

42. Use nodal analysis to find the current through the $-j2$ kΩ capacitor in the circuit of Figure 6.21.
$I1 = 1\text{E}{-3}\angle 0°$, $I2 = 5\text{E}{-3}\angle 0°$, $I3 = 6\text{E}{-3}\angle{-90°}$.

43. Use nodal analysis to find the current through the 3.3 kΩ resistor in the circuit of Figure 6.22.
$E = 36\angle 0°$, $I = 4\text{E}{-3}\angle{-120°}$.

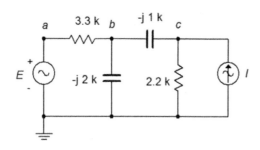

Figure 6.22

44. Given the circuit in Figure 6.22, write the node equations and determine v_c. $E = 18\angle 0°$, $I = 7.5\text{E}{-3}\angle{-30°}$.

45. Given the circuit in Figure 6.23, use nodal analysis to determine v_c. $E = 40\angle 180°$, $I = 20\text{E}{-3}\angle 0°$.

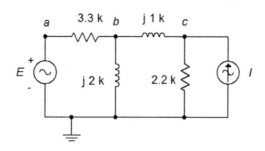

Figure 6.23

46. Use nodal analysis to find the current through the 2.2 kΩ resistor in Figure 6.23. $E = 240\angle 0°$, $I = 100\text{E}{-3}\angle 0°$.

47. Use nodal analysis to find v_{bc} in the circuit of Figure 6.2.

48. Use nodal analysis to find the current through the 2.7 kΩ resistor in the circuit of Figure 6.3.

49. Given the circuit in Figure 6.4, use nodal analysis to determine v_{ba}. $E1 = 1\angle 0°$, $E2 = 2\angle 0°$.

50. Given the circuit in Figure 6.5, use nodal analysis to determine v_{ad}. $E1 = 9\angle 0°$, $E2 = 5\angle 40°$.

51. Use nodal analysis to find v_{cb} in the circuit of Figure 6.7. $E1 = 10\angle{-180°}$, $E2 = 25\angle 0°$.

52. Given the circuit in Figure 6.9, use nodal analysis to determine v_{bc}. $E = 20\angle 0°$, R1 = 10 kΩ, R2 = 30 kΩ, R3 = 1 kΩ, $X_C = -j15$ kΩ, $X_L = j20$ kΩ.

53. Given the circuit in Figure 6.10, use nodal analysis to determine v_{bc}. $E = 120\angle 0°$, R1 = 1 kΩ, R2 = 2 kΩ, R3 = 3 kΩ, $X_C = -j10$ kΩ, $X_L = j20$ kΩ.

54. Determine v_a in the circuit of Figure 6.24 if the source $E = 2\angle 0°$.

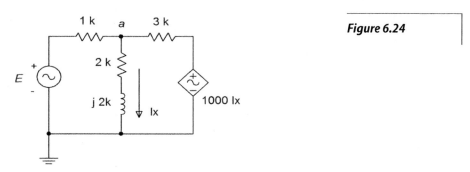

Figure 6.24

55. Given the circuit in Figure 6.24, determine the current flowing through the 1 kΩ resistor. Assume that $E = 15\angle 45°$.

56. Given the circuit in Figure 6.25, determine the current flowing through the 3 kΩ resistor if the source $E = 25\angle 33°$.

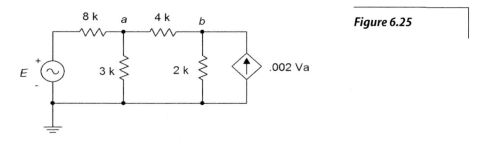

Figure 6.25

57. Given the circuit in Figure 6.25, determine v_{ab}. Assume the source $E = 15\angle -112°$.

58. In the circuit of Figure 6.26, determine v_d.

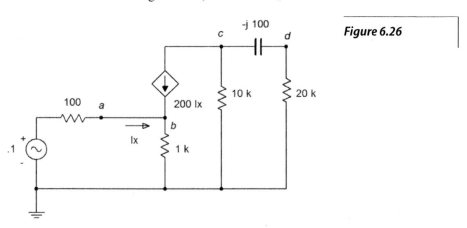

Figure 6.26

59. Given the circuit in Figure 6.26, determine the current flowing through the 1 kΩ resistor.

60. Given the circuit in Figure 6.27, determine the current flowing through the 100 Ω resistor.

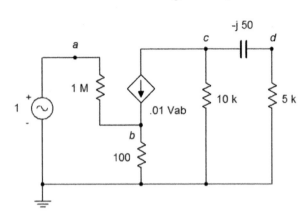

Figure 6.27

61. Determine v_d in the circuit of Figure 6.27.

Challenge

62. Given the circuit in Figure 6.28, write the node equations. $E1 = 50\angle 0°$, $E2 = 35\angle 120°$, $I = 500\text{E}{-}3\angle 90°$.

63. Given the circuit in Figure 6.28, use either mesh or nodal analysis to determine v_{ed}. $E1 = 9\angle 0°$, $E2 = 12\angle 0°$, $I = 50\text{E}{-}3\angle 0°$.

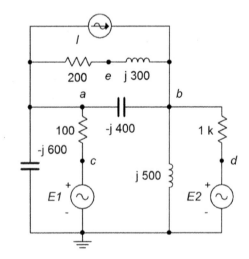

Figure 6.28

64. Given the circuit in Figure 6.29, use mesh analysis to determine v_{fc}. $E1 = 12\angle 0°$, $E2 = 48\angle 0°$, $E3 = 36\angle 70°$.

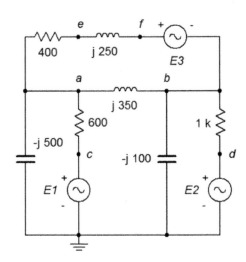

Figure 6.29

65. Find voltage v_{bc} in the circuit of Figure 6.30 using either mesh or nodal analysis. $E = 100\angle 0°$, $R1 = R2 = 2$ kΩ, $R3 = 3$ kΩ, $R4 = 10$ kΩ, $R5 = 5$ kΩ, $X_{C1} = X_{C2} = -j2$ kΩ.

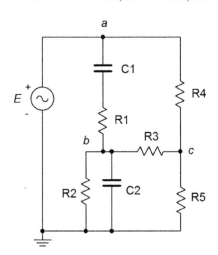

Figure 6.30

66. Given the circuit in Figure 6.31, use nodal analysis to find v_{ac}. $I1 = 8\text{E}{-3}\angle 0°$, $I2 = 12\text{E}{-3}\angle 0°$, $E = 50\angle 0°$.

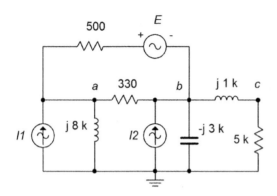

Figure 6.31

67. Given the circuit in Figure 6.32, use nodal analysis to determine v_{ad}. $I1 = 0.1\angle 0°$, $I2 = 0.2\angle 0°$, $I3 = 0.3\angle 0°$.

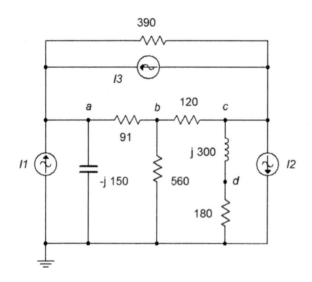

Figure 6.32

68. Given the circuit in Figure 6.33, determine v_{ad}. $E1 = 15\angle 0°$, $E2 = 6\angle 0°$, $I = 100\text{E}{-3}\angle 0°$.

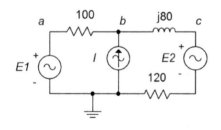

Figure 6.33

69. Given the circuit in Figure 6.34, determine v_{ad}. $E1 = 22\angle 0°$, $E2 = -10\angle 0°$, $I = 2\text{E}{-}3\angle 0°$.

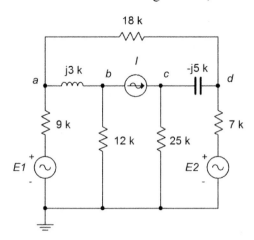

Figure 6.34

70. Given the circuit in Figure 6.35, determine v_{ab}. $I1 = 1.2\angle 0°$, $I2 = 2\angle 120°$, $E = 75\angle 0°$.

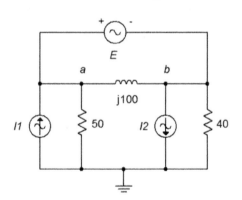

Figure 6.35

71. Given the circuit in Figure 6.36, determine v_{ad}. $I1 = 0.8\angle 0°$, $I2 = 0.2\angle 180°$, $I3 = 0.1\angle 0°$, $E = 15\angle 0°$.

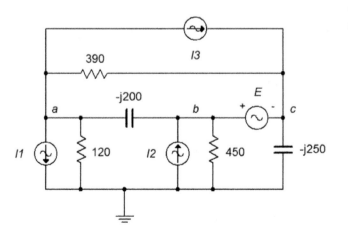

Figure 6.36

Simulation

72. Perform a transient analysis simulation on the circuit of problem 1 (Figure 6.1) to verify the results for v_b.

73. Investigate the variation of v_b due to frequency in problem 1 (Figure 6.1) by performing an AC simulation. Run the simulation from 10 Hz up to 100 kHz.

74. Investigate the variation of v_b due to component tolerance in problem 1 (Figure 6.1) by performing a Monte Carlo simulation. Apply a 10% tolerance to the resistors and capacitor.

75. Perform a transient analysis simulation on the circuit of problem 4 (Figure 6.2) to verify the results for v_c.

76. Investigate the variation of v_b due to frequency in problem 4 (Figure 6.2) by performing an AC simulation. Run the simulation from 1 Hz up to 10 kHz.

77. Investigate the variation of v_b due to component tolerance in problem 4 (Figure 6.2) by performing a Monte Carlo simulation. Apply a 10% tolerance to the resistors and capacitors.

7 AC Power

This section covers:
- Power waveforms.
- Power Triangle.
- Power factor.
- Power factor correction.
- Efficiency.

7.0 Introduction

Power Waveforms

Computation of power in AC systems is somewhat more involved than the DC case due to the phase between the current and voltage. First, consider the case of a purely resistive load, that is, with a phase angle of 0 degrees. To determine the power, simply multiply the voltage by the current. This results in:

$$P(t) = V\sin(2\pi ft) \times I\sin(2\pi ft)$$
$$P(t) = VI\left(\frac{1}{2} - \frac{1}{2}\cos(2\pi 2 ft)\right)$$
$$P(t) = \frac{VI}{2} - \frac{VI}{2}\cos(2\pi 2 ft)$$

The final expression is made of two parts; the first portion which is fixed (not a function of time) and the second portion which consists of a negative cosine wave at twice the frequency. This can be visualized as a time shifted sine wave that is riding on a DC level which is equal to the peak value of the new sinusoid. This is shown in Figure 7A on the following page. Note that green current waveform is drawn just slightly above its true value so that it may be seen easily next to the otherwise identical red voltage waveform. The power product is shown in blue. Due to the fact that sinusoids are symmetrical around zero, the effective power dissipation averaged over time will be the offset value, or $VI/2$.

For example, a one volt peak source delivering a current of one amp peak should generate $VI/2$, or 0.5 watts. This crosschecks nicely with the RMS calculation of roughly 0.707 volts RMS times 0.707 amps RMS also yielding 0.5 watts.

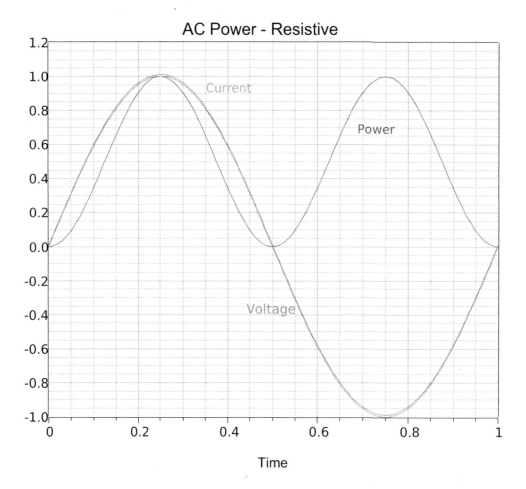

Figure 7A
(current shifted slightly positive for ease of viewing)

The situation is considerably different if the load is purely reactive. For a load consisting of just an inductor, the voltage leads the current by 90 degrees.

$$P(t) = V\cos 2\pi ft \times I \sin 2\pi ft$$
$$P(t) = VI\left(\frac{1}{2}\sin 2\pi 2 ft\right)$$
$$P(t) = \frac{VI}{2}\sin 2\pi 2 ft$$

This is shown in Figure 7B on the following page. In this case, there is no net power dissipation. Power is being alternately generated and dissipated (i.e., positive values indicate dissipation while negative values indicate generation). In this respect, the reactive element can be thought of as alternately storing and releasing energy in the manner of an ideal spring being compressed and then released. The result is essentially the same if the load is purely capacitive.

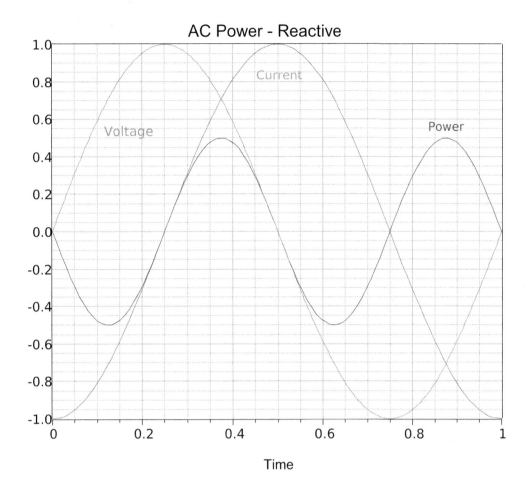

Figure 7B

Finally, we come to the case of a complex load, part resistive and part reactive. Given a phase angle, θ, we have:

$$P(t) = V\sin 2\pi ft \times I \sin 2\pi ft + \theta$$
$$P(t) = VI\left(\frac{1}{2}\cos\theta - \frac{1}{2}\cos 2\pi 2ft + \theta\right)$$
$$P(t) = \frac{VI}{2}\cos\theta - \frac{VI}{2}\cos 2\pi 2ft + \theta$$

These waveforms are shown graphically in Figure 7C on the following page. Note that while this analysis used an inductive load, the same can be said regarding the capacitive case (simply swap the labels for the current and voltage waveforms). As far as the power is concerned, the long term average is now a function of the phase angle, θ. As cos θ may vary between 0 and 1, the power for the complex impedance case will never be more than that of the purely resistive version.

Finally, in the equations above, *V* and *I* are peak values. If RMS values are used, there is no need to divide *VI* by 2.

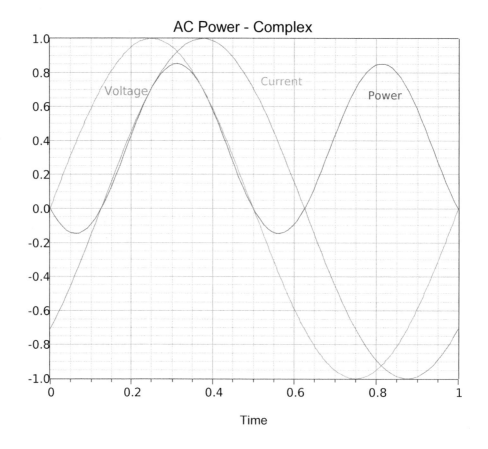

Figure 7C

Power Triangle

A simple calculation of power using the magnitudes of the current and voltage can lead to erroneous results. The phase angle between the current and voltage cannot be ignored. For example, if a 120 volt RMS source delivers 2 amps of current, it appears that it delivers 240 watts. This is only true if the load is purely resistive. For a complex load, the true power is somewhat less. In fact, as we've just seen, if the load is purely reactive, there will be no true load power at all.

To illustrate this, we use a power triangle as shown in Figure 7D. The horizontal axis represents true power, P, in watts. The vertical axis represents "reactive power", Q, or the results we would get if we ignored the phase angle of the reactive component. So that it is not confused with the true power, Q has units of VAR (volt-amps reactive). The vector combination of P and Q results in the *apparent power*, S, which has units of VA (volt-amps). The apparent power is the product of the magnitudes of the current and voltage with the phase angle ignored. This is what the power "appears to be" based on simple current and voltage measurements from a DMM. In the resistive case, there is no reactive power and thus S and P are the same. In a purely reactive case, there is no true power and S and Q are the same. For the complex case, S is the vector sum of P and Q.

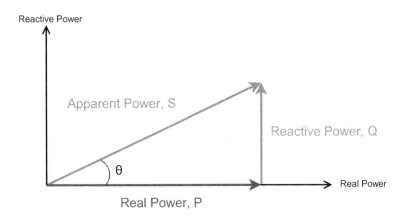

Figure 7D

For example, consider a voltage source delivering 10 volts RMS to a load consisting of a one ohm resistor in series with an inductive reactance of one ohm, $1+j1$. This is equivalent to approximately $1.414\angle 45°$ ohms. The current is $7.07\angle -45°$ amps. The product of the magnitudes is 10 volts times 7.07 amps, or 70.7 VA. This is the apparent power, S. As this is a series connected load, we can use the current to find the power in each element. For the resistor, $P = I^2R$, or 50 watts. The similar calculation for the inductor yields $Q = 50$ VAR. As seen in the prior section, the true power can be determined using the phase angle, $P = VI \cos \theta$ for RMS signals. Thus, $P = 10 \cdot 7.07 \cos 45° = 50$ watts.

Power Factor

Examining Figure 7D shows that $Q = S \sin \theta$ and $P = S \cos \theta$. As we are often interested in the true power, it is worth noting that the ratio of true power to apparent power is the cosine of the impedance angle, $P/S = \cos \theta$. This is known as the *power factor* and is abbreviated PF. Thus, $PF = \cos \theta$. Knowing the phase angle and the apparent power, true power can be calculated. If PF is positive it is said to be a *lagging* power factor. This is the case for inductive loads. A capacitive load results in a negative or *leading* PF. The sign is only used to indicate leading or lagging (useful for power factor correction). For example, if a 100 volt RMS source delivers 1 amp for an apparent power of 100 VA and the phase angle is $-30°$, PF is $\cos(-30°)$ or 0.866 leading and the true power is $P = 100 \cos(-30°) = 86.6$ watts.

Power Factor Correction

One issue with a reactive load is that the current is higher than it needs to be in order to achieve a certain true power. To alleviate this, an opposite reactance can be added to the load such that the resulting load is purely resistive. This can be realized by determining the original Q value and then adding a sufficient reactance to produce an additional Q of the opposite sign, resulting in cancellation. Using the previous example, $Q = 100 \sin(-30°) = 50$ VAR capacitive. Thus, we need to add 50 VAR inductive in parallel with this load. Knowing the applied voltage, the reactance may be determined: $X_L = (100 \text{ volts})^2/50 \text{ VAR} = 200$ ohms. Knowing the frequency, the required inductance can then be found using the inductive reactance formula, $X_L = 2\pi f L$.

Efficiency

Efficiency is defined as usable power output divided by applied power input and is denoted by the Greek letter eta, η. Normally it is expressed as a percentage and it can never be over 100%. Some loads, such as typical heating elements, are modeled as resistances. As such, they can be thought of as being 100% efficient, meaning that all of the electrical input is turned into useful output (in this case, heat). Incandescent light bulbs are also resistive, but they turn most of their input into heat rather than the desired quantity (light) and thus suffer from low efficiency.

In contrast, some loads have a sizable reactive component along with being less than 100% efficient. Examples include motors and loudspeakers. Motors have an inductive component and lose power in the form of mechanical losses (e.g., friction) and electrical losses (e.g., resistance of windings). Motors are rated in terms of their output power, not the power they draw from the source. For example, a motor with a 1 HP rating might be said to generate 1 HP (746 W) *at the shaft*. If the motor is 90% efficient, the electrical draw would be 746 W/.9 or 829 W. This situation is further complicated by the phase angle (i.e., power factor) of the motor due to reactive elements, as noted in the previous sections. Compared to motors, home loudspeakers are particularly inefficient, typically converting only about 1% of the electrical input into usable acoustic output. The impedance of a moving coil dynamic loudspeaker is usually inductive although it can be capacitive in some parts of the frequency spectrum. The less common electrostatic loudspeaker presents a highly capacitive load and also exhibits low efficiency.

7.1 Exercises

Analysis

1. For the circuit shown in Figure 7.1, determine apparent power S, real power P, reactive power Q and power factor PF. Also, draw the power triangle.

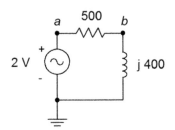

Figure 7.1

2. For the circuit shown in Figure 7.2, determine apparent power S, real power P, reactive power Q and power factor PF. Also, draw the power triangle.

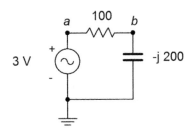

Figure 7.2

3. For the circuit shown in Figure 7.3, determine apparent power S, real power P, reactive power Q and power factor PF. Also, draw the power triangle. The source is 120 volts.

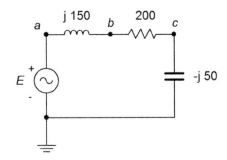

Figure 7.3

4. For the circuit shown in Figure 7.4, determine apparent power S, real power P, reactive power Q and power factor PF. Also, draw the power triangle. The source is 120 volts.

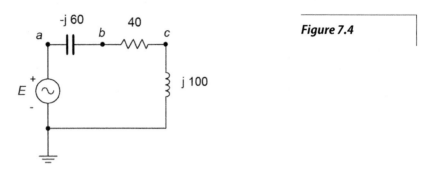

Figure 7.4

5. For the circuit shown in Figure 7.5, determine apparent power S, real power P, reactive power Q and power factor PF. The source is 90 volts, $X_L = j30$ Ω, R = 50 Ω.

Figure 7.5

6. For the circuit shown in Figure 7.6, determine apparent power S, real power P, reactive power Q and power factor PF. Also, draw the power triangle. The source is 240 volts, $X_C = -j200$ Ω, R = 75 Ω.

Figure 7.6

7. For the circuit shown in Figure 7.7, determine apparent power S, real power P, reactive power Q and power factor PF. The source is 120 volts, $X_L = j40$ Ω, $X_C = -j25$ Ω, R = 20 Ω.

Figure 7.7

8. For the circuit shown in Figure 7.7, determine apparent power S, real power P, reactive power Q and power factor PF. The source is 120 volts, 60 Hz. R = 80 Ω, C = 20 µF, L = 400 mH.

9. An audio power amplifier delivers a 30 volt RMS 1 kHz sine to a loudspeaker. If the loudspeaker impedance at this frequency is 7∠45°, determine the RMS current delivered to the load and the true power.

10. An audio power amplifier delivers an 80 volt peak 35 Hz sine to a subwoofer. If the subwoofer impedance at this frequency is 4∠−30°, determine the peak current delivered to the load and the true power.

11. A certain load is specified as drawing 8 kVA with a lagging power factor of 0.8. Determine the real power P, and the reactive power Q. Further, if the source is 120 volts at 60 Hz, determine the effective impedance of the load in both polar and rectangular form, and the requisite resistance/inductance/capacitance values.

12. A certain load is specified as drawing 20 kVA with a leading power factor of 0.9. Determine the real power P, the reactive power Q and draw the power triangle. If the source is 240 volts at 60 Hz, determine the effective impedance of the load in both polar and rectangular form, and the requisite resistance/inductance/capacitance values.

13. Consider the system shown in Figure 7.8. E is a standard 120 V input. If the three loads are 45 W, 60 W and 75 W incandescent light bulbs, respectively, determine the apparent power delivered to the system, the source current, the reactive power and the real power.

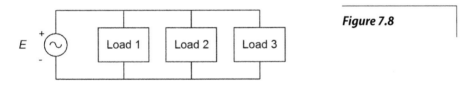

Figure 7.8

14. Given the system shown in Figure 7.8, determine the apparent power delivered to the system, the source current, the real power, the reactive power and the efficiency. E is 120 V. The three loads are resistive heating elements of 500 W, 1200 W and 1500 W, respectively.

15. Consider the system shown in Figure 7.9. E is 240 V. If the three loads are 200 W, 400 W and 1000 W resistive, respectively, determine the apparent power delivered to the system, the real power and the reactive power.

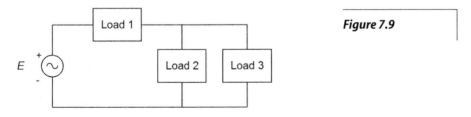

Figure 7.9

16. Given the system shown in Figure 7.9, determine the apparent power delivered to the system, the real power, the reactive power and the efficiency. E is 480 V. The three loads are resistive heating elements of 1500 W, 2000 W and 3500 W, respectively.

17. Consider the system shown in Figure 7.8. E is 120 V. Load 1 is 1 kW resistive, load 2 is 400 W resistive and load 3 is 600 VAR inductive. Determine the apparent power delivered to the system, the source current, the reactive power, the real power and the power factor.

18. Consider the system shown in Figure 7.8. E is 240 V. Load 1 is 2 kW resistive, load 2 is 800 W resistive and load 3 is 1200 VAR capacitive. Determine the apparent power delivered to the system, the source current, the real power, the reactive power and the power factor.

19. Given the system shown in Figure 7.8, determine the apparent power delivered to the system, the source current, the real power, the reactive power and the power factor. E is 120 V. Load 1 is 600 W of incandescent lighting, load 2 is 1200 W of heating elements and load 3 is 200 VAR capacitive.

20. Given the system shown in Figure 7.8, determine the apparent power delivered to the system, the source current, the real power, the reactive power and the power factor. E is 60 V. Load 1 is 90 W of incandescent lighting, load 2 is 800 W of heating elements from a dryer and load 3 200 VAR inductive.

21. A 120 V 3 HP motor draws a real power of 2500 W from the source. Determine its efficiency.

22. A 120 V 12 HP motor draws a real power of 10 kW from the source. Determine its efficiency.

23. An ideal 120 V 2 HP motor draws an apparent power of 1800 W from the source. Determine its power factor.

24. An ideal 120 V 0.3 HP motor draws an apparent power of 270 W from the source. Determine its power factor.

25. A 120 V motor is rated at 0.5 HP. It has an efficiency of 78% and a lagging power factor of 0.7. Determine the apparent power drawn from the source (S), the real power (P), and the reactive power (Q) supplied. Also draw the power triangle and find the delivered current.

26. A motor is rated at 10 HP. It has an efficiency of 92% and a lagging power factor of 0.8. Determine the apparent power drawn from the source (S), the real power (P), and the reactive power (Q) supplied. Also draw the power triangle. Finally, determined the current drawn from the 120 V source.

27. Consider the system shown in Figure 7.8. E is 120 V. Load 1 is 1 kW resistive, load 2 is 400 W resistive and load 3 is a 1 HP motor that is 80% efficient and has a 0.85 lagging power factor. For the system, determine the apparent power delivered, the source current, the real power, the reactive power and the power factor.

28. Consider the system shown in Figure 7.8. E is 120 V. Load 1 is 2.5 kW resistive, load 2 is 500 VAR capacitive and load 3 is a 2 HP motor that is 85% efficient and has a 0.9 lagging power factor. For the system, determine the total power delivered, the source current, the apparent power, the real power and the power factor.

29. For the system shown in Figure 7.10, E is 240 V. Load 1 is 1.2 kW resistive heating, load 2 is 400 W resistive lighting, load 3 is a 0.5 HP motor that is 80% efficient with a 0.7 lagging power factor, and load 4 is a 1 HP motor that is 85% efficient with a 0.8 lagging power factor. For the system, determine the apparent power delivered, the source current, the real power, the reactive power and the power factor.

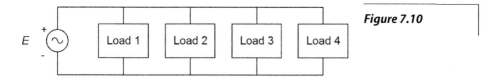

Figure 7.10

Design

30. A 120 V 60 Hz source drives a load equivalent to a 75 Ω resistor in parallel with a 25 µF capacitor. Determine the appropriate capacitance or inductance value to place across this load to produce unity power factor.

31. A 240 V 60 Hz source drives a load equivalent to a 10 Ω resistor in parallel with a 50 mH inductor. Determine the appropriate capacitance or inductance value to place across this load to produce unity power factor.

32. A load of 50∠30° is driven by a 120 V 60 Hz source. Determine the appropriate capacitance or inductance value to place across this load to produce unity power factor.

33. A load of 50∠−50° is driven by a 240 V 60 Hz source. Determine the appropriate capacitance or inductance value to place across this load to produce unity power factor.

34. A certain load is specified as drawing 8 kVA with a lagging power factor of 0.8. The source is 120 volts at 60 Hz. Determine the appropriate capacitor or inductor to place in parallel with this load to produce unity power factor.

35. A certain load is specified as drawing 20 kVA with a leading power factor of 0.9. The source is 240 volts at 60 Hz. Determine the appropriate capacitor or inductor to place in parallel with this load to produce unity power factor.

36. A 240 V 60 Hz source is connected to a load consisting of heating elements of 10 kW along with a 15 HP motor with η=90%, PF=0.85. Determine an appropriate capacitor or inductor to place in parallel to produce unity power factor.

37. A 120 V 60 Hz source is connected to a load consisting of 350 W of resistive lighting along with a 1.5 HP motor with η=70%, PF=0.75. Determine an appropriate capacitor or inductor to place in parallel to produce unity power factor.

Challenge

38. A power distribution system for a concert has the following specifications: Ten class D audio power amplifiers rated at 2 kW output each with 90% efficiency and unity power factor, 10 kW worth of resistive stage lighting to illuminate the musicians alongside a troupe of trained dancing kangaroos, a 3 HP motor used to continuously rotate the drum riser throughout the performance (η=80%, PF=0.75) and a 2 HP compressor which inflates and deflates a giant rubber *T. rex* during particularly exciting parts of the show (η=85%, PF=0.8). For the system, determine the total power delivered, the source current, the apparent power, the real power, and the power factor. Finally, make a sketch of this extravaganza with its entertainers in full regalia singing their latest tune "Maximum Volume".

Simulation

39. Verify the design of problem 28 by performing a transient analysis. The design will have been successful if the source current and voltage are in phase.

40. Verify the design of problem 29 by performing a transient analysis. The design will have been successful if the source current and voltage are in phase.

8 Resonance

This section covers:
- Series resonance.
- Parallel resonance.

8.0 Introduction

Series Resonance

Resonance can be thought of as a *preferred frequency of vibration*. It is exploited in a variety of areas, for example, a good mechanical resonance can be used for the construction of acoustic musical instruments. For a series RLC circuit, the resonant frequency, f_0, is the frequency at which the magnitudes of X_C and X_L are the same. If the capacitive and inductive reactance formulas are set equal to each other and simplified, we arrive at a formula for the resonant frequency:

$$f_0 = \frac{1}{2\pi\sqrt{LC}}$$

Note that a particular resonant frequency can be obtained through a variety of LC pairs. A simple series resonant circuit is shown in Figure 8A.

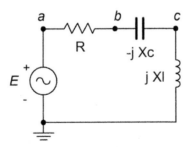

Figure 8A

The combined impedance at any frequency is $R + jX_L - jX_C$. At very low frequencies, X_C dominates and the circuit impedance approaches the purely capacitive case. At very high frequencies, X_L dominates and the circuit appears inductive. At resonance, X_C and X_L cancel leaving just R. This implies that power factor is unity at resonance. In a real world circuit, R is the combination of the series resistance plus any resistance from the inductor's coil. This is illustrated in Figure 8B, following.

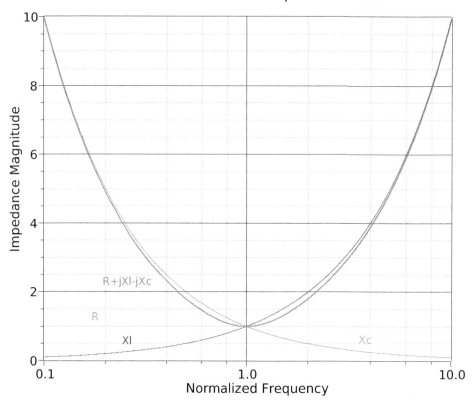

Figure 8B

The corresponding phase plot is shown in Figure 8C. Note the capacitive phase angle at low frequencies and the inductive phase angle at high frequencies. Further, at resonance, the phase angle is zero, showing that the circuit is purely resistive. Of particular importance is the circuit Q, which is the ratio of the reactance to the total resistance at resonance. Q will create a multiplying effect on the inductor and capacitor voltages at resonance.

$Q_{series} = X_0/R$

Where R is the total series resistance ($R_{series}+R_{coil}$).

At f_0, the current through the circuit will equal the source voltage divided by R (because X_C and X_L cancel). This current is also flowing through the capacitor and inductor. Because their reactances are Q times higher than R, then their voltages will be Q times higher than the source voltage. KVL is not violated because, again, the voltages across L and C are 180 degrees out of phase and cancel each other.

As the circuit Q is increased, the voltage multiplying effect becomes more pronounced. Also, the impedance curve tends to become sharper near resonance and the phase change becomes more abrupt. This is shown in Figure 8D.

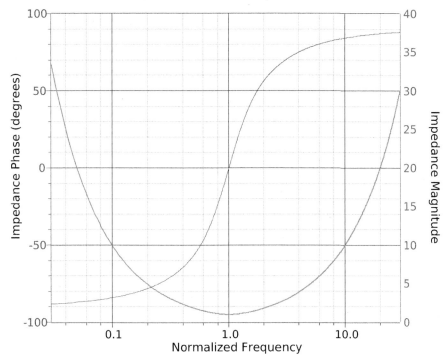

Figure 8C

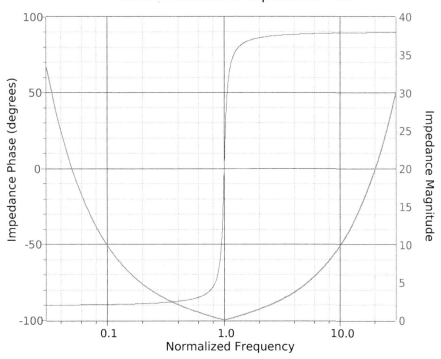

Figure 8D

The "sharpness" of the curve is related to the half power or "−3 dB" frequencies, f_1 and f_2. These are the frequencies at which the current (assuming voltage source drive) falls off to 0.707 of the maximum value at resonance. f_1 is below f_0 and f_2 is found above. The difference between these two frequencies is called the bandwidth, BW.

$$BW = f_2 - f_1$$
$$BW = f_0/Q_{series}$$

In general, the ratio f_0/f_1 equals f_2/f_0. For higher Q circuits ($Q \geq 10$), we can approximate symmetry, and thus

$$f_1 = f_0 - BW/2$$
$$f_2 = f_0 + BW/2$$

As an example, consider a circuit with the following parameters: 10 volt peak source, L=1 mH, C=1 nF, and a total circuit resistance (i.e., R_{series}+ R_{coil}) of R = 50 Ω.

$$f_0 = \frac{1}{2\pi\sqrt{LC}}$$
$$f_0 = \frac{1}{2\pi\sqrt{1e\text{-}3 \cdot 1e\text{-}9}}$$
$$f_0 = 159 \text{ kHz}$$

$X_L = 2\pi f L = 2\pi 159 \text{ kHz } 1 \text{ mH} = 1000 \text{ } \Omega$
$Q_{series} = X_L/R = 1000/50 = 20$
$BW = f_0/Q_{series} = 159 \text{ kHz}/20 = 7.95 \text{ kHz}$
$f_1 \approx 155 \text{ kHz}$
$f_2 \approx 163 \text{ kHz}$

Given the 10 volt peak source, the voltages across the capacitor and inductor at the resonance frequency of 159 kHz would be Q times greater, or 200 volts. At higher or lower frequencies, the increased impedance lowers the current and also lowers the voltages across the components. At low frequencies, most of the source will appear across the capacitor while at high frequencies the inductor voltage will approach the source voltage.

A practical point regarding R_{coil} and Q_{coil}: While it is possible to measure the DC resistance of a coil using a DMM, this will not necessarily give an accurate value at high frequencies. Thus, a preferred method is to determine Q_{coil} at the desired frequency from the inductor's spec sheet, and using the calculated reactance at that frequency, determine the value of R_{coil}.

Parallel Resonance

If the three RLC components are placed in parallel, parallel resonance will be of interest. Parallel resonance is slightly more complicated than series resonance due to the fact that the series coil resistance cannot be lumped in with the remaining circuit resistance as it can with the series case. To alleviate this problem, it is possible to find a parallel equivalent for the series inductive reactance and associated coil resistance.

Series to Parallel Inductor Transform

Assume a practical coil exists as a series combination of resistance and inductive reactance, $R_s + jX_s$. The parallel equivalent, $R_p \| jX_p$, is found as follows. Begin with the reciprocal conductance/resistance rule:

$$R_s + jX_s = \frac{1}{\dfrac{1}{R_p} + \dfrac{1}{jX_p}}$$

$$\frac{1}{R_s + jX_s} = \frac{1}{R_p} + \frac{1}{jX_p}$$

Using the complex conjugate, isolate the real and imaginary parts of the series version,

$$\frac{1}{R_s + jX_s} \frac{R_s - jX_s}{R_s - jX_s} = \frac{R_s}{R_s^2 + X_s^2} + \frac{-jX_s}{R_s^2 + X_s^2}$$

Therefore,

$$\frac{1}{R_p} = \frac{R_s}{R_s^2 + X_s^2}$$

$$\frac{1}{jX_p} = \frac{-jX_s}{R_s^2 + X_s^2}$$

And,

$$R_p = \frac{R_s^2 + X_s^2}{R_s}$$

$$jX_p = j\frac{R_s^2 + X_s^2}{X_s}$$

If $X_s \gg R_s$ then we can approximate these as

$$R_p \approx \frac{X_s^2}{R_s} = Q_{coil} X_s = Q_{coil}^2 R_s$$

$$jX_p \approx j\frac{X_s^2}{X_s} = jX_s$$

Thus, the parallel equivalent reactance is unchanged from the series value and the parallel equivalent resistance is the series resistance times the Q of the coil squared (recalling that $Q_{coil} = X_L/R_{coil}$).

For higher Q circuits ($Q \geq 10$), the resonant frequency is found as it is in the series case:

$$f_0 = \frac{1}{2\pi\sqrt{LC}}$$

For lower Q circuits, the resonant frequency will be reduced slightly due to the fact that the transformed resistance is frequency dependent. Also, due to the "inversion" of the series-parallel transform (i.e., a smaller R_S resulting in a larger R_P), parallel Q is found as:

$$Q_{parallel} = R_P/X_L$$

A parallel impedance plot is shown in Figure 8E. The effect is the inverse of the series case. At low frequencies, the small inductive reactance dominates resulting in a low impedance magnitude with a positive (inductive) phase angle. At high frequencies, the small capacitive reactance dominates resulting in a low impedance magnitude with a negative (capacitive) phase angle. At resonance, the reactive values cancel leaving just the parallel resistive value which produces the characteristic peak in impedance with a phase angle of zero.

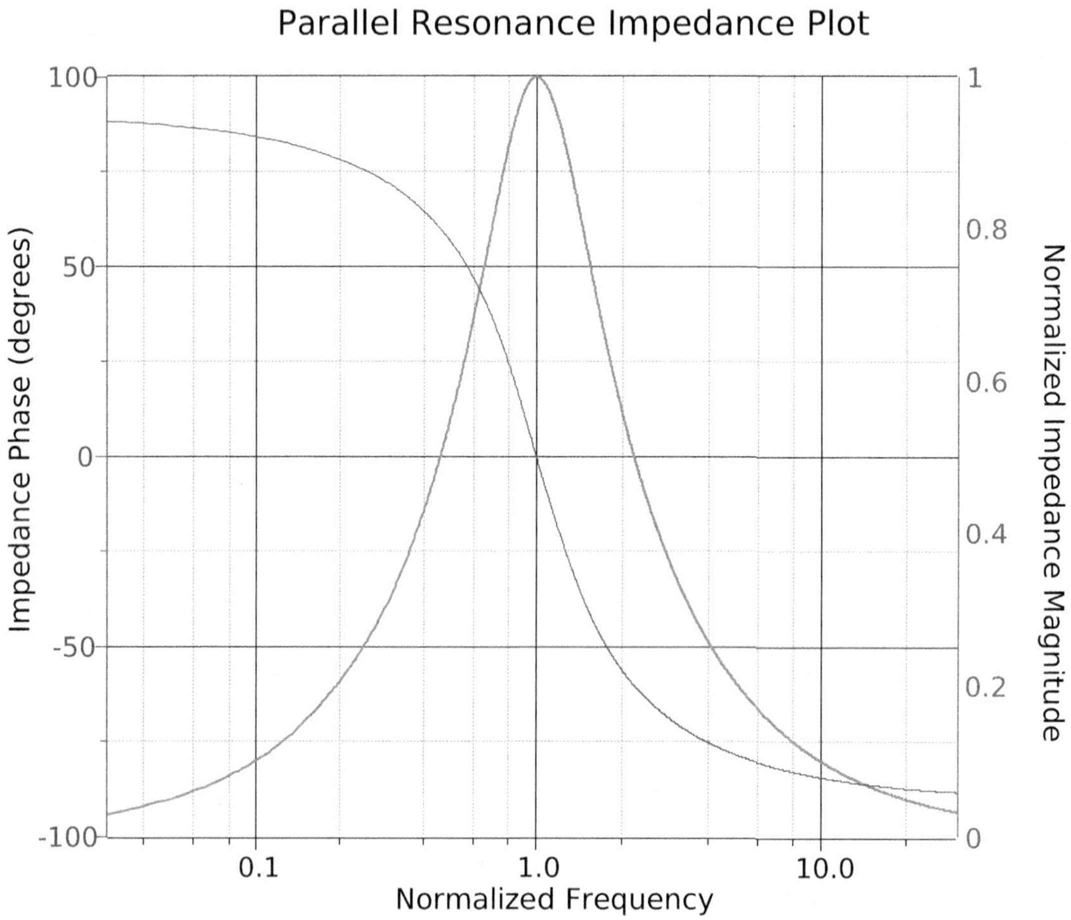

Figure 8E

If the parallel resonant circuit is driven by a current source, then the voltage produced across the resonant circuit (sometimes referred to as a *tank circuit*) will echo the shape of the impedance magnitude. In other words, it will effectively discriminate against high and low frequencies, and keep only those signals in the vicinity of the resonant frequency. This is one method of making a *bandpass filter*. The lower and upper half power frequencies, f_1 and f_2, are found in the same manner as in series resonance.

Repeating for convenience:

$$BW = f_2 - f_1$$
$$BW = f_0/Q_{parallel}$$

In general, the ratio f_0/f_1 equals f_2/f_0. For higher Q circuits ($Q \geq 10$), we can approximate symmetry, and thus

$$f_1 = f_0 - BW/2$$
$$f_2 = f_0 + BW/2$$

As the parallel Q increases, the impedance curve becomes sharper and the phase change is more abrupt, as in the series case. A comparison of high and low Q curves for parallel resonance is shown in Figure 8F.

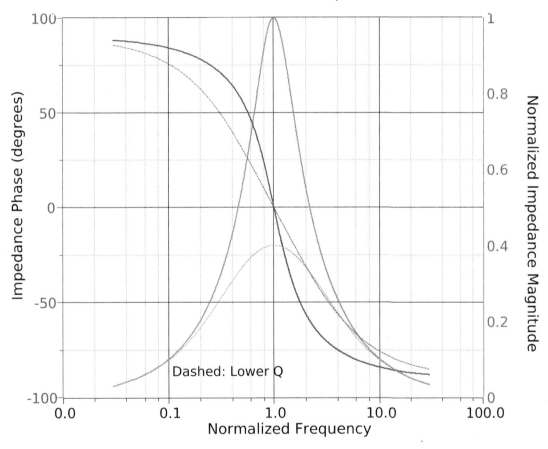

Figure 8F

As an example, consider a circuit with the following parameters: 1 mA peak source, L=2 mH, C=10 nF, R = 10 kΩ, and Q_{coil} = 25.

$$f_0 = \frac{1}{2\pi\sqrt{LC}}$$
$$f_0 = \frac{1}{2\pi\sqrt{2e\text{-}3 \cdot 1e\text{-}8}}$$
$$f_0 = 35.59 \text{ kHz}$$

$X_L = 2\pi f L = 2\pi 35.59 \text{ kHz } 2 \text{ mH} = 447 \ \Omega$
$R_{coil} = X_L/Q_{coil} = 447/25 = 17.9 \ \Omega$

The parallel equivalent of the coil resistance is

$R_P = R_{coil} \cdot Q^2_{coil} = 17.9 \cdot 25^2 = 11.18 \text{ k}\Omega$

This is in parallel with R yielding an effective parallel resistance of 11.18 kΩ || 10 kΩ, or 5.28 kΩ.

$Q_{parallel} = R_P/X_L = 5.28 \text{ k}/447 = 11.07$
$BW = f_0/Q_{parallel} = 35.59 \text{ kHz}/11.8 = 3.016 \text{ kHz}$
$f_1 \approx 34.1 \text{ kHz}$
$f_2 \approx 37.1 \text{ kHz}$

With a 1 mA source, the voltage at the resonant frequency will be approximately 5.28 volts. The voltage will drop off on either side of 35.59 kHz.

As with series resonance, there is an apparent "Q amplification" effect in parallel resonant circuits, however, here it will be the reactive currents that will be increased relative to the source current instead of the component voltages.

Note that the parallel resistor can be used to lower the system Q and thus broaden the bandwidth, however, the system Q can never be higher than the Q of the inductor itself. The inductor sets the upper limit on system Q and therefore, how tight the bandwidth can be.

Finally, it is worth repeating that for relatively low Q values there will be some shifting of the resonant and half power frequencies from the equations presented above.

8.1 Exercises

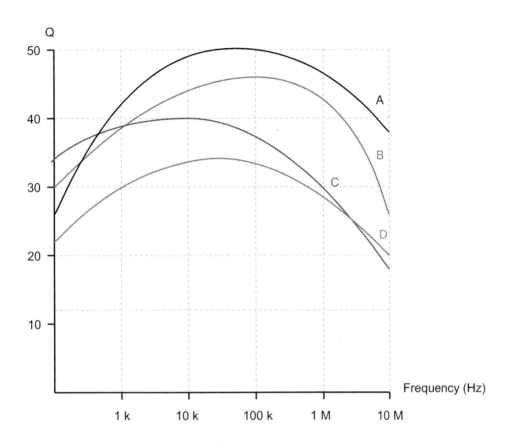

Inductor Q curves to be used with the exercises below

Analysis

1. A circuit has a resonant frequency of 440 kHz and a system Q of 30. Determine the bandwidth and the approximate values for f_1 and f_2.

2. A circuit has a resonant frequency of 19 kHz and a bandwidth of 500 Hz. Determine the system Q and the approximate values for f_1 and f_2.

3. Find the Q_{coil} and coil resistance of a 150 µH inductor at 100 kHz using device curve A.

4. Find the Q_{coil} and coil resistance of a 2.2 mH inductor at 50 kHz using device curve D.

5. At a certain frequency, an inductor's impedance is $24 + j600\ \Omega$. Determine the parallel resistance and reactance that produces the same value.

6. At a certain frequency, an inductor's impedance is $3 + j150\ \Omega$. Determine the parallel resistance and reactance that produces the same value.

7. A certain 75 µH inductor is described by curve B. Determine the equivalent parallel inductor/resistor combination at 1 MHz.

8. A certain 3.3 mH inductor is described by curve C. Determine the equivalent parallel inductor/resistor combination at 20 kHz.

9. Consider a series circuit consisting of a 2 nF capacitor, an ideal 33 µH inductor and a 5 Ω resistor. Determine the resonant frequency, system Q, and bandwidth.

10. Consider a series circuit consisting of a 20 nF capacitor, an ideal 100 µH inductor and a 2.7 Ω resistor. Determine the resonant frequency, system Q, and bandwidth.

11. Consider a series circuit consisting of a 50 nF capacitor, a 20 mH inductor with Q_{coil} of 50 and a 63 Ω resistor. Determine the resonant frequency, system Q, and bandwidth.

12. Consider a series circuit consisting of a 200 nF capacitor, a 1 mH inductor with Q_{coil} of 65 and a 72 Ω resistor. Determine the resonant frequency, system Q, and bandwidth.

13. For the circuit shown in Figure 8.1, determine the resonant frequency, system Q and bandwidth. Assume $R_{coil} = 0$ Ω. If the source is 1 volt peak, determine the capacitor voltage at resonance.

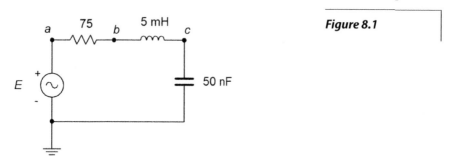

Figure 8.1

14. For the circuit shown in Figure 8.2, determine the resonant frequency, system Q and bandwidth. Assume $R_{coil} = 0$ Ω. If the source is 10 volts, determine the capacitor voltage at resonance.

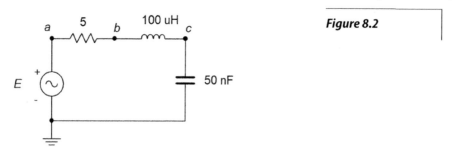

Figure 8.2

15. Repeat problem 13 but assume instead that the inductor's $R_{coil} = 15$ Ω.

16. Repeat problem 12 but assume instead that the inductor follows curve D.

17. For the circuit shown in Figure 8.3, determine the resonant frequency, system Q and bandwidth. If the source is 20 mA peak, determine the resistor and capacitor voltages at resonance.

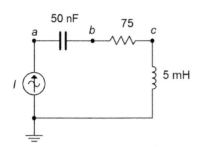

Figure 8.3

18. For the circuit shown in Figure 8.4, determine the resonant frequency, system Q and bandwidth. If the source is 100 mA, determine the resistor and capacitor voltages at resonance.

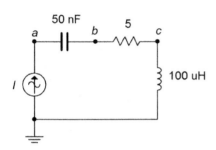

Figure 8.4

19. For the circuit shown in Figure 8.5, determine the resonant frequency, system Q and bandwidth. If the source is 15 volts, determine the inductor and capacitor currents at resonance. Assume the inductor's coil resistance is 3.2 Ω.

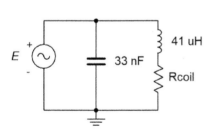

Figure 8.5

20. For the circuit shown in Figure 8.6, determine the resonant frequency, system Q and bandwidth. If the source is 3 volts, determine the inductor and capacitor currents at resonance. Assume the inductor's Q is 30.

Figure 8.6

21. For the circuit shown in Figure 8.7, determine the resonant frequency, system Q and bandwidth. If the source is 15 volts, determine the resistor, inductor and capacitor currents at resonance.

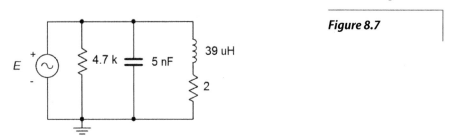

Figure 8.7

22. Given the circuit shown in Figure 8.8, determine the resonant frequency, system Q and bandwidth. If the source is 2 volts, determine the resistor, inductor and capacitor currents at resonance. Assume the inductor's coil resistance is 2.5 Ω.

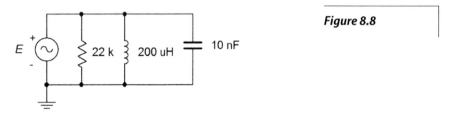

Figure 8.8

23. For the circuit shown in Figure 8.9, determine the resonant frequency, system Q and bandwidth. If the source is 5 volts, determine the resistor, inductor and capacitor currents at resonance. Assume the inductor's Q is 40.

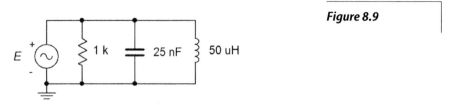

Figure 8.9

24. Given the circuit shown in Figure 8.10, determine the resonant frequency, system Q and bandwidth. If the source is 2.5 mA, determine the resistor voltage and the three branch currents at resonance.

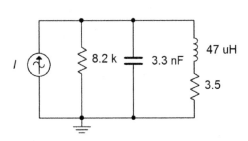

Figure 8.10

25. For the circuit shown in Figure 8.11, determine the resonant frequency, system Q and bandwidth. If the source is 500 µA, determine the resistor voltage and the three branch currents at resonance. Assume the inductor's Q is given by curve C.

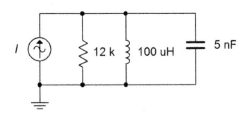

Figure 8.11

26. Given the circuit shown in Figure 8.12, determine the resonant frequency, system Q and bandwidth. If the source is 10 mA, determine the resistor voltage and the three branch currents at resonance. Assume the inductor's Q is given by curve B.

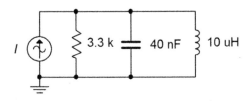

Figure 8.12

Design

27. A series resonant circuit has a required f_o of 50 kHz. If a 75 nF capacitor is used, determine the required inductance.

28. A series resonant circuit has a required f_o of 210 kHz. If a 22 µH inductor is used, determine the required capacitance.

29. A parallel resonant circuit consists of a 12 nF capacitor and a 27 µH inductor with a Q_{coil} of 55. Determine the required additional parallel resistance to achieve a system Q of 40.

30. A series resonant circuit has a design target of f_0=200 kHz with a bandwidth of 5 kHz. Which of the inductor curves above (A, B, C, D) represent possible candidates, if any, and why/why not?

31. A parallel resonant circuit has a design target of f_0=1 MHz with a bandwidth of 20 kHz. Which of the inductor curves above (A, B, C, D) represent possible candidates, if any, and why/why not?

Challenge

32. A parallel resonant circuit has a required f_0 of 50 kHz and a bandwidth of 4 kHz. If a 75 nF capacitor is used and the load impedance is 100 kΩ, determine the required inductance and minimum acceptable Q_{coil}.

33. A parallel resonant circuit consists of a 150 nF capacitor and a 200 μH inductor that has a coil resistance of 1 Ω. The desired bandwidth for the network is 2 kHz. Determine the value of resistance to be placed in parallel with the network in order to achieve this goal.

34. A resonant circuit consists of a 4 nF capacitor in parallel with a 100 μH coil that has a coil resistance of 5 Ω. Determine the resonant frequency and bandwidth. Further, assume that this circuit is now loaded by an amplifier that has an input impedance equivalent to 10 kΩ resistive in parallel with 500 pF of input capacitance. Also, the amplifier is connected via 25 feet of coaxial cable that exhibits a capacitance of 33 pF per foot. Determine the changes in resonant frequency and bandwidth, if any, with this load.

Simulation

35. Use an AC frequency domain analysis to verify the results of problem 13. Plot the resistor voltage from $0.1\, f_0$ to $10\, f_0$.

36. Use an AC frequency domain analysis to verify the results of problem 19. Do this by overlapping plots of the resistor, capacitor and inductor voltages across a range of $0.1\, f_0$ to $10\, f_0$.

37. Investigate the effects of inductor Q on the system bandwidth of problem 21. Plot the system voltage from $0.01\, f_0$ to $100\, f_0$ three times, the first using the specified coil resistance and then using values ten times larger and ten times smaller.

38. Investigate the effects of component tolerance on the system frequency response of problem 21. Plot the system voltage from $0.1\, f_0$ to $10\, f_0$ using a Monte Carlo variation on the AC frequency domain response. Set a 10% tolerance on the capacitor, inductor and resistor but do not alter the coil resistance.

39. Use an AC frequency domain analysis to verify the design of problem 27. Plot the resistor voltage from $0.1\, f_0$ to $10\, f_0$.

126

40. Use an AC frequency domain analysis to verify the design of problem 29. Plot the system voltage from $0.1 f_0$ to $10 f_0$.

41. At high Q values (>10) the capacitor and inductor voltages of series resonant circuits will tend to reach maximum very close to the resonant frequency. At lower Qs, these peaks tend to diverge. A similar situation occurs with the currents in parallel resonant circuits. Investigate this effect by performing an AC frequency domain analysis on problem 14. Overlay plots of v_{ab}, v_{bc} and v_c for successively larger values of resistance.

42. Investigate the "Q increase" in reactive currents compared to source and resistive currents in a parallel resonant circuit. A simple way to verify this is by placing AC ammeters in each of the branches of the circuit shown in Figure 8.13. Use R = 630 Ω, C = 40 nF, L = 10 μH and I = 1 mA. It is worthwhile to compare sets of simulations for different resistor values to see the current changes relative to the system Q. Slight variations of the source frequency may be required to reach the peak.

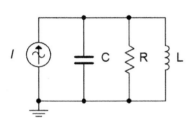

Figure 8.13

9 Polyphase Power

This section covers:
- Three phase systems in both Δ (delta) and Y (wye).

9.0 Introduction

Polyphase Definition

A polyphase system uses multiple current-carrying wires with multiple generators, each with their own unique phase. This allows for considerable delivery of power to the load. The most popular scheme is the three-phase configuration. This can be visualized as three individual sine generators that are interconnected, as shown in Figures 9A and 9B. Figure 9A shows a Δ (i.e., *delta* and also known as π) connected system while Figure 9B shows a Y (also known as *wye* or *T*) connected system (here drawn upside down so that nodes *a*, *b* and *c* match locations with Figure 9A).

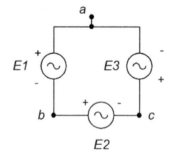

Figure 9A

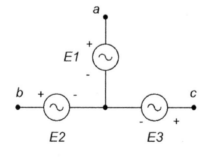

Figure 9B

Of particular importance is the relative phase of each source. As the load will also have three segments or *legs* (a three phase load), a consistent delivery of power demands that the three sources be spread equally over time. This means that each source is one-third of a cycle, or 120 degrees, out of phase with the other legs (i.e., leading one and lagging the other). This is shown in Figure 9C. We shall only consider the case of *balanced loads*, that is, where each leg of the load is identical to the other legs.

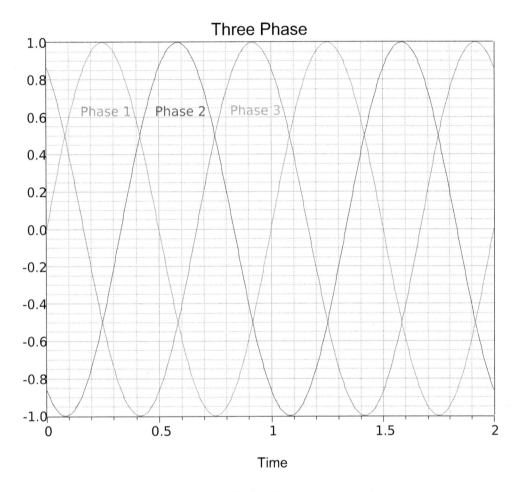

Figure 9C

It is possible to configure systems using delta or Y connected sources with either delta or Y connected loads. One item to note is that delta connected systems are always three wire systems (nodes a, b and c as drawn above) while Y connected systems can make use of a fourth wire (the common point that all three sources connect to). The most straightforward systems are delta-to-delta and Y-to-Y. These are shown in Figures 9D and 9E, respectively, following.

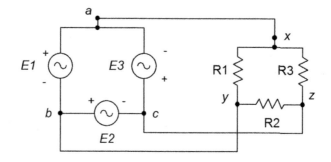

Figure 9D

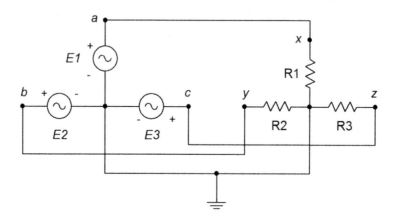

Figure 9E

For a delta connected source, the generator voltage is equal to the *line voltage* (that is, the voltage as measured between any pair of the three external nodes: a, b and c). Thus, in Figure 9D, it should be obvious that $V_{ab} = E1$. This is also the voltage developed across each of the legs of the load, in this example, R1 (i.e., V_{xy}). The situation is different for the Y connected source. In Figure 9E, the generator voltage (e.g., $E1$) is developed from an external node to the common point rather than from node to node. A vector summation shows that the line voltage is √3 times the generator voltage. Thus, if three 120 volt sources are connected in a Y configuration, the line voltage will be √3 times 120 volts, or approximately 208 volts. V_{ab}, V_{bc} and V_{ca} will all have a magnitude of 208 volts, each 120 degrees out of phase with the others. Further, note that $V_{ab} = V_{xy}$ and also $E1 = V_{R1}$.

A similar situation occurs regarding line versus generator current. In a Y connected system, it should be obvious that the generator current must be equal to the line current (e.g., the current flowing from node a to the load, node x). This is also the load current. In contrast, for a delta connected system, the currents of two generator legs combine to form the line current. A vector summation shows that the line current must be √3 times the generator current. The current in each leg of the load will equal the current in each leg of the source (i.e., generator current). Alternately, the load current must equal the line current divided by √3.

For mixed systems, the situation is slightly more complex. Y-to-delta and delta-to-Y schemes are shown in Figures 9F and 9G, respectively.

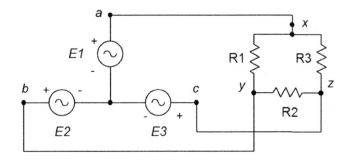

Figure 9F

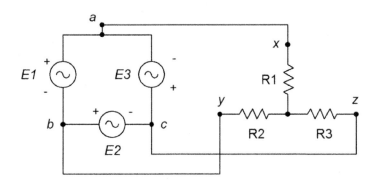

Figure 9G

In Figure 9F, the line voltage equals √3 times the generator voltage. The load is delta connected, so each leg sees the line voltage. Based on this, each leg of the load current can be computed. Note that the line current equals the generator current. The load current will be the line current divided by √3.

In Figure 9G, the line voltage equals the generator voltage. The load is Y connected, so each leg sees the line voltage divided by √3. Based on this, each leg of the load current can be computed. Note that the line current equals the load current. The generator current will be the line current divided by √3.

The preceding analysis only examines one leg of the three, thus the total power generated or delivered will be three times that of a single leg. Also, although resistive loads are shown, reactive loads are common. The analysis is similar to the above but with consideration of the phase angles. Finally, note that it is possible to have a load that is both delta and Y connected. For example, an inductive load such as a motor might have its three legs configured in a Y arrangement while three capacitors are connected in a delta arrangement for power factor correction.

9.1 Exercises

Unless specified otherwise, assume generator frequencies are 60 Hz for all problems.

Analysis

1. As depicted in Figure 9.1, a 3-phase Δ connected generator feeds a Δ connected load. The generator phase voltage is 120 volts and the load consists of 3 legs of 10 Ω each. Find the voltage across each load leg, the line current through the wires connecting the load to to the generator and the power drawn by the load.

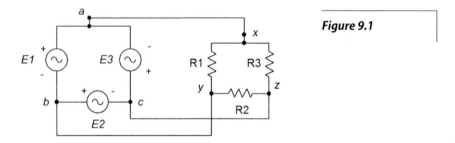

Figure 9.1

2. Referring to the delta-delta system of Figure 9.1, if the generator phase voltage is 230 volts and the load is balanced with each leg at 2 Ω, determine the line voltage, line current, generator phase current and load current.

3. The system of Figure 9.2 shows a 3-phase Y connected generator feeding a Y connected load. If the generator phase voltage is 120 volts and the load consists of 3 legs of 20 Ω each, find the line voltage, the line current, voltage across each load leg and the total power drawn by the load.

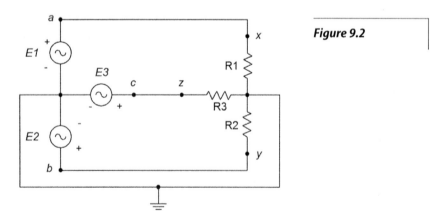

Figure 9.2

4. Referring to Figure 9.2, if the generator phase voltage is 230 volts and the load is balanced with each leg at 12 Ω, determine the line voltage, line current, generator phase current, load current, load voltage and total load power.

132

5. As depicted in Figure 9.3, a 3-phase Δ connected generator feeds a Y connected load. The generator phase voltage is 120 volts and the load consists of balanced legs of 5 Ω each. Find the voltage across each load leg, the line current, the line voltage, the generator phase current and the total load power.

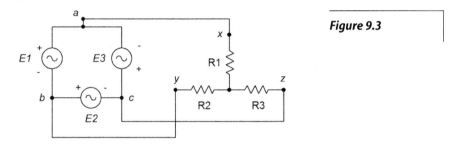

Figure 9.3

6. Referring to Figure 9.3, if the generator phase voltage is 400 volts and the load is balanced with each leg at 10 Ω, determine the line voltage, line current, generator phase current, load current and the voltage across each load leg.

7. The system of Figure 9.4 shows a 3-phase Y connected generator feeding a Δ connected load. If the generator phase voltage is 120 volts and the load consists of 3 legs of 60 Ω each, find the line voltage, the line current, voltage across each load leg and the total power drawn by the load.

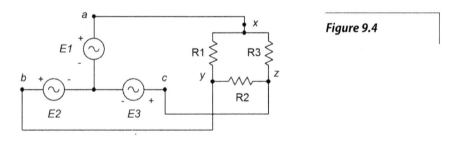

Figure 9.4

8. Referring to the wye-delta system of Figure 9.3, if the generator phase voltage is 120 volts and the load is balanced with each leg at 20 Ω, determine the line voltage, line current, generator phase current, load current, the voltage across each load leg and the total load power.

9. A 3-phase Δ connected generator feeds a Δ connected load consisting of 3 legs of 10 Ω in series with $j4$ Ω of inductive reactance, as shown in Figure 9.5. If the line voltage is 208 volts, find the voltage across each load leg, the current through the wires connecting the load to to the generator, and the apparent and real powers drawn by the load.

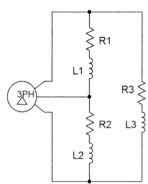

Figure 9.5

10. Given the delta-delta system of Figure 9.5, if the generator phase voltage is 120 volts and the load is balanced with each leg at $20 + j10$ Ω, determine the line voltage, line current, generator phase current, load current, the voltage across each load leg, and the total real and apparent load powers.

11. A 3-phase Y connected generator feeds a Y connected load consisting of 3 legs of 10 Ω in series with $j4$ Ω of inductive reactance, as shown in Figure 9.6. If the line voltage is 208 volts, find the voltage across each load leg, the line current, and the apparent and real powers drawn by the load.

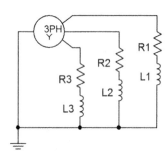

Figure 9.6

12. Given the wye-wye system of Figure 9.6, if the line voltage is 400 volts and the load is balanced with each leg at $100 + j20$ Ω, determine the generator phase voltage, line current, generator phase current, load current, the voltage across each load leg, and the total real and apparent load powers.

13. The 3-phase system of Figure 9.7 uses a Y connected generator feeding a Δ connected load. The load consists of 3 legs of 40 Ω in series with $j30$ Ω of inductive reactance, as shown in Figure 9.7. If the generator phase voltage is 230 volts, find the line voltage, the voltage across each load leg, the line current, the load current, and the apparent and real powers drawn by the load.

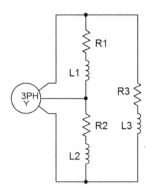

Figure 9.7

14. Given the wye-delta system of Figure 9.7, if the line voltage is 400 volts and the load is balanced with each leg at $80 + j20$ Ω, determine the generator phase voltage, line current, generator phase current, load current, the voltage across each load leg, and the total real and apparent load powers.

15. A 208 3-phase Δ connected generator feeds a Y connected load consisting of 3 legs of 10 Ω in series with $j4$ Ω of inductive reactance as shown in Figure 9.8. Find the voltage across each load leg, the current through the wires connecting the load to to the generator, and the apparent and real powers drawn by the load.

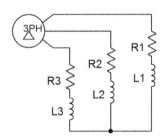

Figure 9.8

16. Given the delta-wye system of Figure 9.8, if the line voltage is 400 volts and the load is balanced with each leg at $120 + j30$ Ω, determine the line current, generator phase current, load current, the voltage across each load leg, and the total real and apparent load powers.

17. A 120 volt 3-phase Δ connected generator feeds a Δ connected load consisting of 3 legs of 75 Ω in series with $-j10$ Ω of capacitive reactance as shown in Figure 9.9. Find the voltage across each load leg, the current through the wires connecting the load to to the generator, and the apparent and real powers drawn by the load.

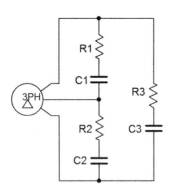

Figure 9.9

18. A 3-phase Y connected generator feeds a Y connected load consisting of 3 legs of 150 Ω in series with $-j20$ Ω of capacitive reactance as shown in Figure 9.10. If the generator phase voltage is 120 volts, find the line voltage, the voltage across each load leg, the line current, and the apparent and real powers drawn by the load.

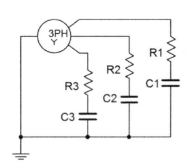

Figure 9.10

Design

19. Using the delta-delta system of problem 9 and assuming the source frequency is 60 Hz, determine appropriate component values to place in parallel with each load leg in order to shift the power factor to unity.

20. Using the Y-Y system of problem 11 and assuming the source frequency is 60 Hz, determine appropriate component values to place in parallel with each load leg in order to shift the power factor to unity.

Challenge

21. Using the Y-Y system of problem 11 and assuming the source frequency is 60 Hz, determine appropriate component values to be added to the load in order to shift the power factor to unity. These new components should be in a delta configuration.

Simulation

22. Use a transient analysis to verify the phase and line voltage phase relationships in problem 1.

23. Use a transient analysis to verify the results computed for problem 15.

24. Use a transient analysis to verify the design solution to problem 19. This can be achieved by ensuring that the voltage and current in each load leg (with added correction components) are in phase.

25. Use a transient analysis to verify the design solution to problem 20. This can be achieved by ensuring that the voltage and current in each load leg (with added correction components) are in phase.

Notes

10 Magnetic Circuits and Transformers

This section covers:
- Basic magnetic circuits using *B-h* curves.
- Basic transformer operation.

10.0 Introduction

Magnetic Circuits

Magnetic circuits include applications such as transformers and relays. A very simple magnetic circuit is shown in Figure 10A. It consists of a magnetic core material. The core may be comprised of a single material such as sheet steel but can also use multiple sections and air gap(s). Around the core is at least one set of turns of wire. Multiple sets of turns are used for transformers (in the simplest case, one for the primary and another for the secondary). Passing current through the windings generates a magnetic flux, Φ, in the core. As this flux is constrained with the cross sectional area of the core, A, we can derive a flux density, B, from $B = \Phi/A$. Given the characteristics of the core material, the flux density gives rise to a magnetizing force, H.

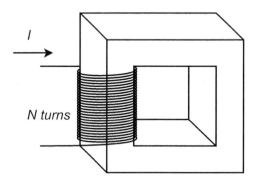

Figure 10A

The flux density and magnifying force are related through a *BH* curve. An example of *BH* curves for three different core materials is shown in Figure 10B[1]. Knowing the length of the core, the H*l* product can be found.

---

1 Curves based on https://en.wikipedia.org/wiki/Saturation_(magnetic) and Boylestad, Introductory Circuit Analysis, 12E. A: sheet steel, B: cast steel, C: cast iron.

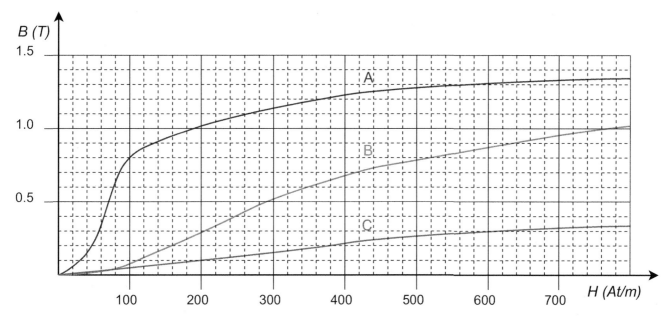

Figure 10B

In a magnetic circuit, the current times the number of turns, or *NI*, can be seen as analogous to a voltage rise. Similarly, the *Hl* product is analogous to a voltage drop. Thus, the sum of the *Hl* drops must equal the sum of the *NI* rises. In the circuit of Figure 10A, there is a single "rise" and a single "drop". The core could consist of two or more different materials, creating the equivalent of a series circuit. In this case, a table such as the one found in Figure 10C can be used to aid in computation.

Section	Flux Φ (Wb)	Area A (m^2)	Flux Density B (T)	Magnetizing Force H (At/m)	Length l (m)	Hl (At)
1						
2						

Figure 10C

For example, suppose we have a core made of two materials and we would like to determine the current required to obtain a certain flux. Given the physical dimensions of the core, we can quickly fill in the values for the area and length of the two sections. As these sections are in series, the flux must be the same for each of them (i.e, the stated desired level). At this point we can now determine the value for the flux density. The left half of the table is now filled. Having computed the flux density, we can jump over to right side by using the *BH* curves for the two sections. Multiplying these values by their lengths yield the associated *Hl* "drops". The sum of these two must equal the *NI* "rise". Knowing the number of turns, the current can be determined.

If an air gap is used, the value for *H* can be found using the relation: $H \approx 8E5\ B$.

Transformers

Transformers have three basic uses: 1) isolation, 2) voltage scaling, and 3) impedance matching. Transformers exploit magnetic circuit theory for their operation. Typically, the device has two windings, one for the input side (the primary) and one for the output side (the secondary), although it is possible to have sets of input and/or output windings. A typical schematic symbol is shown in Figure 10B. As far as the construction is concerned, picture the diagram of Figure 10A but with an added second winding on the right side serving as the secondary.

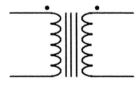

Figure 10B

Of particular importance is the ratio of the number of turns on the primary to the number of turns on the secondary, or N (referred to as the *turns ratio*). As the two windings are on the same core and ideally see the same flux, the voltage ratio of the primary to secondary will be equal to N for an ideal (lossless) transformer. Thus, if $N=10$ and the primary voltage is 60 volts, then the secondary voltage will be 6 volts. The primary and secondary appear as "NI" potentials in the magnetic circuit, therefore, the current will change inversely with N. For example, if $N=10$ and the primary current is 1 amp then the secondary current will be 10 amps. It is important to note that the product of primary current and voltage must equal the product of secondary current and voltage for an ideal transformer. Such a device does not dissipate power but rather *transforms* it (hence the name). Real world devices will exhibit some loss due to finite winding wire resistance, non-ideal behavior of the magnetic circuit and the like. Instead of a power dissipation rating, power transformers are given a VA (volt-amps) rating, typically thought of as secondary voltage times maximum secondary current.

Because there is no direct electrical connection between the primary and secondary, the transformer may be used solely for *isolation*. This is used for safety. If $N=1$, then there is no change in the voltage being presented to the device (ideally). In contrast, power applications will make use of N to scale the voltage to a more appropriate level. If the voltage is brought down ($N>1$), it is referred to as a *step-down transformer*. If $N<1$, it is referred to as a *step-up transformer*. For example, a consumer electronics product may require a modest DC voltage for its operation, say 15 or 20 volts. To achieve this, the standard North American wall voltage of 120 volts may be reduced using a step-down transformer with $N=5$. This would result in a secondary voltage of 24 volts that could then be rectified, filtered, and regulated to the desired DC level. Conversely, a step-up transformer could be used to create a much higher secondary potential (for example, for long distance transmission). For some applications, split secondaries or multiple secondaries may be used. A center-tap is relatively common and it splits the secondary into two equal voltage sections (i.e., a 24 volt CT would behave as two 12 volt secondaries in series). A transformer with a center tapped secondary is shown in Figure 10C, following. Split secondaries are also common and are more flexible than a center-tapped secondary. Split secondaries can be combined in series to increase the secondary voltage or combined in parallel to increase secondary current capacity (i.e, load current).

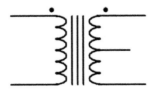

Figure 10C

In order to determine the phase relationship between the primary and secondary, transformers use "dot notation". The dot indicates the positive polarity of voltage. Refer to the circuit of Figure 10D.

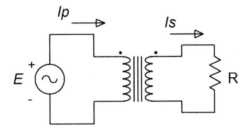

Figure 10D

On the primary side, the current flows into the dot and establishes the positive reference. In this case, the primary is seen as the load for the source, E. On the secondary side, current flows out of the dot and also notates the positive reference because the secondary is seen as the source for the load, R.

The third use of the transformer is impedance matching. Because both the voltage and current are being scaled, the source's "view" of the load changes. In fact, as the voltage and current are moving in opposite directions by N, then the load impedance changes by N^2. For example, suppose source E in Figure 10D is generating 20 volts, $N=5$ and the load R is 8 Ω. The secondary voltage is $V_p/N = 20$ volts/5 = 4 volts. Therefore, $I_s = 4$ volts/8 Ω = 0.5 amps. The current will scale with $1/N$, so this means that the primary current, I_p, must equal $I_s/N = 0.5$ amps/5 = 0.1 amps. Before continuing, note that the product of secondary voltage and current equals the product of primary voltage and current, as expected (2 volt-amps). Now, if we consider the impedance that the source E drives, that is found via Ohm's law, $V_p/I_p = 20$ volts/0.1 amps = 200 Ω. This is the same as the load resistance multiplied by N^2 (8 $\Omega \cdot 5^2 = 200$ Ω). This is called the *reflected impedance*. If the source, E, had a large internal resistance compared to the load, R, and there was no transformer, there would be considerable loss in the system. For example, if the source impedance was 10 Ω, a voltage divider between it and the 4 Ω load would cause a great deal of internal power dissipation in the source leading to inefficiency. On the other hand, the given circuit with transformer reflects an impedance of 200 Ω, and this will cause very little loss compared to the 10 Ω internal resistance of the source yielding a much more efficient transfer.

10.1 Exercises

Analysis

1. In the magnetic circuit shown in Figure 10.1, assume the cross section is 1 cm by 1 cm with a path length of 8 cm. The entire core is made of sheet steel and there are 100 turns on the winding. Determine the current to establish a flux of 8E−5 webers.

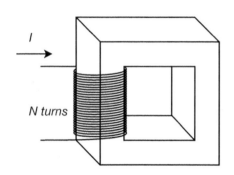

Figure 10.1

2. Repeat problem 1 using cast steel for the core.

3. Given the core shown in Figure 10.1, assume the cross section is 2 cm by 2 cm with a path length of 10 cm. The entire core is made of cast steel and there are 200 turns on the winding. Determine the current to establish a flux of 4E−4 webers.

4. Repeat problem 3 using sheet steel for the core.

5. In the magnetic circuit shown in Figure 10.1, assume the cross section is 1 cm by 1 cm with a path length of 8 cm. The entire core is made of sheet steel. Determine the number of turns required to establish a flux of 8E−5 webers given a current of 50 mA.

6. Given the core shown in Figure 10.1, assume the cross section is 2 cm by 2 cm with a path length of 10 cm. The entire core is made of sheet steel Determine the number of turns required to establish a flux of 4E−4 webers given a current of 200 mA.

7. In the magnetic circuit shown in Figure 10.2, assume the cross section is 1 cm by 1 cm. Section A is sheet steel with a path length of 6 cm. Section B is cast steel with a length of 2 cm. There are 500 turns on the winding. Determine the current to establish a flux of 5E−5 webers.

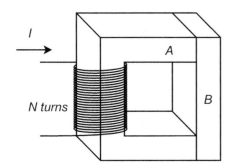

Figure 10.2

8. In the magnetic circuit shown in Figure 10.2, assume the cross section is 1 cm by 1 cm. Section A is sheet steel with a path length of 6 cm. Section B is cast steel with a length of 2 cm. Determine the number of turns on the coil to establish a flux of 6E−5 webers with a current of 50 mA.

9. A transformer is shown in Figure 10.3, assume the cross section is 5 cm by 5 cm. The core is sheet steel with a path length of 20 cm. $N1$ is 500 turns and $N2$ is 200 turns. Determine the secondary current ($I2$) if a primary current ($I1$) of 1 amp establishes a flux of 1.5E−3 webers.

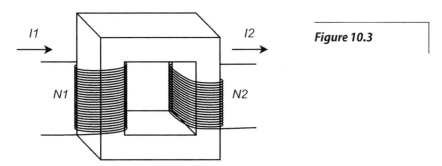

Figure 10.3

10. Given the same conditions of problem 9, alter the secondary turns ($N2$) so that the secondary current ($I2$) is 3 amps given the original primary current of 1 amp.

11. In general, how would the performance noted in problem 9 change if cast steel was substituted for sheet steel?

12. Given the results of problems 9 through 11, what does the ratio of $N1$ to $N2$ represent in terms of idealized performance, and what steps should be taken to make the transformer operate as close to ideal as possible?

13. Given the magnetic circuit shown in Figure 10.4, assume the cross section is 1 cm by 2 cm with a path length of 6 cm. The entire core is made of sheet steel with the exception of a 1 mm air gap. Determine the current required to establish a flux of 4E−4 webers if $N = 1000$ turns.

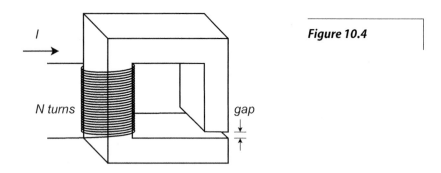

Figure 10.4

14. Using the data given in problem 13, determine the number of turns required to establish the same flux when the input current is 200 mA.

15. An ideal transformer has a 6:1 voltage step-down ratio. If the primary is driven by 24 VAC and the load is 100 Ω, determine the load voltage and current, and the primary side current.

16. A 120 VAC transformer is specified as having a 36 volt center-tapped secondary. If each side of the secondary is connected to its own 50 Ω load, determine the load currents and the primary side current.

17. An ideal transformer has a 12:1 voltage step-down ratio. If the secondary is connected to a 10 Ω load, what impedance is seen from the primary side?

18. An ideal transformer has a 1:5 voltage step-up ratio. If the secondary is connected to a 2 Ω load, what impedance is seen from the primary side?

Design

19. A step-down transformer with $N=8$ has a 15 volt RMS secondary which is connected to a load with an effective value of 5 Ω. Determine the minimum acceptable VA rating of the transformer.

20. A step-up transformer with $N=0.5$ is driven from a 120 VAC source. The secondary is connected to a load with an effective value of 150 Ω. Determine the minimum acceptable VA rating of the transformer.

Challenge

21. A transformer specified as having a 120 VAC primary with an 18 volt secondary is accidentally connected backwards, with its secondary connected to the source and its primary connected to a 16 Ω load. Determine the load current in both the normal and reversed connections. Also determine the required transformer VA rating for both connections.

22. Consider the distributed public address system for an airport as shown in Figure 10.5. It consists of an audio power amplifier with a nominal 70 volt RMS output that is connected to four remote loudspeakers, each separate from the others and some 150 meters away from the amplifier. Each loudspeaker assembly includes a 10:1 voltage step-down transformer that feeds the loudspeaker impedance of 8 Ω (resistive) off its secondary. These four lines are fed in parallel by the amplifier. Determine the power delivered to each loudspeaker and the total current delivered by the power amplifier. Assume the transformers are ideal and ignore any cable resistance.

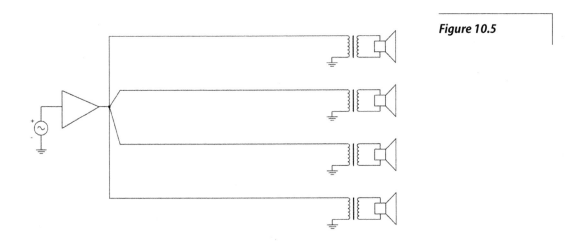

Figure 10.5

23. Continuing with the preceding problem, assume that the wiring connecting each transformer back to the amplifier is AWG 22. Determine the power lost in each of the 150 meter long sections of dual cable. Further, suppose the system is reconfigured without the transformers and the output of the amplifier is lowered to 7 volts RMS to compensate. Determine the power lost in each of the cable feeds under the new configuration.

Simulation

24. Use a transient analysis to verify the results of problem 15.
25. Use a transient analysis to verify the results of problem 16.

Appendix A

Standard Component Sizes

Passive components (resistors, capacitors and inductors) are available in standard sizes. The tables below are for resistors. The same digits are used in subsequent decades up to at least 1 Meg ohm (higher decades are not shown). Capacitors and inductors are generally not available in as many standard values as are resistors. Capacitors below 10 nF (.01 µF) are usually available at the 5% standard digits while larger capacitances tend to be available at the 20% standards.

5% and 10% standard values, EIA E24 and EIA E12

10% values (EIA E12) are **bold**
20% values (seldom used) are every fourth value starting from 10
(i.e., every other 10% value)

10	11	**12**	13	**15**	16	**18**	20	**22**	24	**27**	30
33	36	**39**	43	**47**	51	**56**	62	**68**	75	**82**	91

1% and 2% standard values, EIA E96 and EIA E48

2% values (EIA E48) are **bold**

10.0	10.2	**10.5**	10.7	**11.0**	11.3	**11.5**	11.8	**12.1**	12.4	**12.7**	13.0
13.3	13.7	**14.0**	14.3	**14.7**	15.0	**15.4**	15.8	**16.2**	16.5	**16.9**	17.4
17.8	18.2	**18.7**	19.1	**19.6**	20.0	**20.5**	21.0	**21.5**	22.1	**22.6**	23.2
23.7	24.3	**24.9**	25.5	**26.1**	26.7	**27.4**	28.0	**28.7**	29.4	**30.1**	30.9
31.6	32.4	**33.2**	34.0	**34.8**	35.7	**36.5**	37.4	**38.3**	39.2	**40.2**	41.2
42.2	43.2	**44.2**	45.3	**46.4**	47.5	**48.7**	49.9	**51.1**	52.3	**53.6**	54.9
56.2	57.6	**59.0**	60.4	**61.9**	63.4	**64.9**	66.5	**68.1**	69.8	**71.5**	73.2
75.0	76.8	**78.7**	80.6	**82.5**	84.5	**86.6**	88.7	**90.9**	93.1	**95.3**	97.6

Appendix B

Answers to Selected Numbered Problems

1 Fundamentals

1. 10, 7.07, 0, 1 kHz, 1 ms, 0°

3. 20, −3, 50 Hz, 20 ms, 0°

5. 10, 7.07, 0, 100 Hz, 10 ms, 45°

7. 1, 10, 400 Hz, 2.5 ms, −45°

9. 200 µs, 10 µs

11. 36°

13. 14.14∠45°, 11.2∠−63.4°, 102∠169°, 5k∠53.1°

15. 7.07+j7.07, j.4, −4.5+j7.79, 70.7−j70.7

17. 15+j30, j4, −20−j4, −70+j250

19. −34.5k+j36k, −725+j95, 2.39−j0.709, −2.71−j.457

21. 100∠0°, 10∠−115°, 0.5∠145°, 0.25∠−45°

23. 2.7∠180°, 4.91∠−92.7°, 0.076∠123°, 544∠4.5°

25. −j15.9 k, −j318, −j15.9, −j0.398, −j15.9E−3 Ω

27. −j318 M, −j6.77 M, −j144.7 k, −j96.5 Ω

29. j6.28, j314, j6.28 k, j251 k, j6.28 M Ω

31. j62.8, j3.14 k, j62.8E−3, j2.51 Ω

33.

35. 1.67 @ 3 kHz, 1 @ 5 kHz, 0.714 @ 7 kHz, 0.555 @ 9 kHz, 0.455 @ 11 kHz

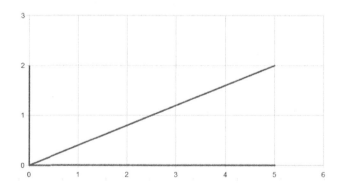

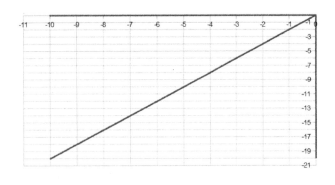

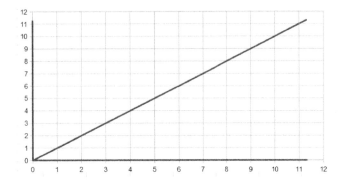

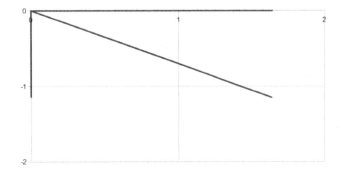

37. 284 nF, 482 pF, 339 pF, 132.6 nF, 212 nF

39. 892 nH, 525 μH, 748 μH, 1.91 μH, 1.19 μH

41. a

43. b

2 Series RLC Circuits

1. $2k - j1.94 k \Omega$

3. $270 + j125.7 \Omega$

5. $1 k - j1.278 k \Omega$

7. $300 - j400$

9. $1.447 \mu F$

11. $v(t) = 0.1\sin 2\pi 1000t$ (i and v are in phase)

13. i is 241 mA p-p and lags by 90°

15. v is 16.6 V p-p and leads by 90°

17. $1 k - j318 \Omega$

19. $i = 953E-6\angle 17.6°$ amps, $v_R = 953E-3\angle 17.6°$ volts, $v_C = 303E-3\angle -72.4°$ volts, delay = 25 µs

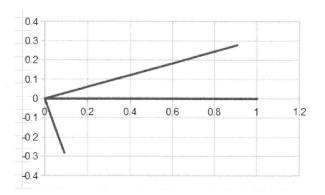

21. $1 k + j3.14 k\Omega$

23. $1 k + j942 \Omega$

25. $2 k - j33.4 \Omega$

27. $i = 0.4999E-3\angle .957°$ amps, $v_R = 0.998\angle .957°$ volts, $v_C = 16.7E-3\angle -89°$ volts, delay = 167 µs

29. $i = 21.2E-3\angle 45°$ amps, $v_R = 4.24\angle 45°$ volts, $v_C = 4.24\angle -45°$ volts

31. $v_S = 60.9\angle -23.2°$ volts, $v_R = 56\angle 0°$ volts, $v_C = 24\angle -90°$ volts

33. $i = 329E-6\angle 70.8°$ amps, $v_R = 329E-3\angle 70.8°$ volts, $v_C = 1.05\angle -19.2°$ volts, $v_L = 103E-3\angle 160.8°$ volts

35. $i = 48.7E-3\angle -13°$ amps, $v_R = 9.745\angle -13°$ volts, $v_C = 3.88\angle -103°$ volts, $v_L = 6.125\angle 77°$ volts

37. $i = 493\text{E}{-}3\angle 9.52°$ amps, $v_B = 58.2\angle 41.6°$ volts, $v_{AC} = 63\angle -29°$ volts

39. $i = 1\text{E}{-}3\angle 0°$ amps, $v_R = 1\angle 0°$ volts, $v_C = 200\text{E}{-}3\angle -90°$ volts, $v_L = 200\text{E}{-}3\angle 90°$ volts

41. $i = 2.24\text{E}{-}3\angle 63.4°$ amps, $v_B = 8.94\angle -26.6°$ volts, $v_C = 2.24\angle 153.4°$ volts, $v_{AC} = 12\angle -4.8°$ volts

43. $i = 35.36\text{E}{-}3\angle -45°$ amps, $v_C = 2.12\angle -135°$ volts, $v_R = 1.414\angle -45°$ volts, $v_L = 3.54\angle 45°$ volts

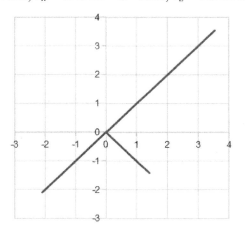

45. $v_R = 10\angle 0°$ volts, $v_C = 5.3\angle -90°$ volts, $v_L = 7.07\angle 90°$ volts

47. $v_R = 2.4\angle 0°$ volts, $v_L = 85.4\text{E}{-}3\angle 90°$ volts, $v_C = 424\text{E}{-}3\angle 90°$ volts

49. $v_R = 110\angle 0°$ volts, $v_C = 88\angle -90°$ volts, $v_L = 220\angle 90°$ volts

51. $v_{AC} = 60\angle 87.8°$ volts, $v_B = 20.1\angle -83.3°$ volts, $v_C = 20\angle -90°$ volts

53. $v_R = 0.329\angle 70.8°$ volts, $v_L = 0.1034\angle 160.8°$ volts, $v_C = 1.048\angle -19.2°$ volts

55. $L = 79.6\ \mu\text{H}$, $C = 7.24$ nF

57. $v_R = 7.23\angle -130°$ volts, $v_L = 3.62\angle -40.1°$ volts, $v_C = 1.45\angle 139.9°$ volts

59. $f = 15.594$ kHz

61. $f = 3.185$ kHz

3 Parallel RLC Circuits

1. $73.2\angle-12.5°$ (71.4 $-j15.8$)

3. $99\angle-8.04°$ (98 $-j13.8$)

5. $182.9\angle52.4°$ (111.5 $+j145$)

7. $f = 45.2$ MHz

9.
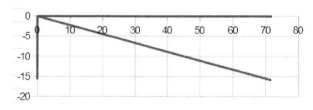

11. $i_s = 0.120001\angle.18°$, $i_R = 0.12\angle0°$, $i_C = 377\text{E}{-}6\angle90°$

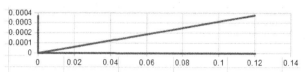

13. $i_s = 9.43\text{E}{-}3\angle-58°$, $i_R = 5\text{E}{-}3\angle0°$, $i_C = 2\text{E}{-}3\angle90°$, $i_L = 10\text{E}{-}3\angle-90°$

15. $i_s = 47.4\text{E}{-}3\angle16.3°$, $i_R = 45.5\text{E}{-}3\angle0°$, $i_C = 20\text{E}{-}3\angle90°$, $i_L = 6.667\text{E}{-}3\angle-90°$ (all peak)

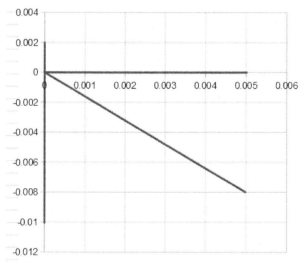

17. $i_R = 19.99\text{E}{-}3\angle-1.73°$, $i_C = 603\text{E}{-}6\angle88.27°$

19. $v_R = v_L = 627\text{E}{-}3\angle87°$ volts

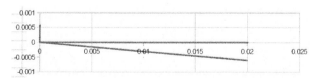

21. $v_s = 2.82\angle-28.1°$ volts

23. $i_{2.2k} = 5.06\text{E}{-}3\angle-68.2°$, $i_{4.7k} = 2.37\text{E}{-}3\angle-68.2°$, $i_C = 55.7\text{E}{-}3\angle21.8°$, $i_L = 37.1\text{E}{-}3\angle-158.2°$

25. $i_C = 670\text{E}{-}6\angle 150°$, $i_L = 536\text{E}{-}6\angle{-}30°$, $i_R = 2.23\text{E}{-}3\angle 60°$

27. $i_R = 1.67\text{E}{-}3\angle{-}5°$, $i_C = 835\text{E}{-}6\angle 85°$, $i_L = 1.044\text{E}{-}3\angle{-}95°$

29. 2.84 µF

31. 7.7 nF

33. 34.5 nF

35. 65.7 nF

37. 390 nF

39. 507 nF

41. 128 mH

4 Series-Parallel RLC Circuits

1. $Z_{10k} \approx 125\angle 0°$, $Z_{1M} \approx 125\angle{-}0.8°$, $Z_{100M} \approx 68.8\angle{-}24.2°\ \Omega$

3. $Z_{300} \approx 0.75\angle 90°$, $Z_{30k} \approx 77.5\angle 89.4°$, $Z_{3M} \approx 907\angle 5.4°\ \Omega$

5. $Z = 180\angle 51.6°\ \Omega$

7. $i_s = 9.64\text{E}{-}3\angle{-}85.6°$, $i_R = 740.7\text{E}{-}6\angle 0°$, $i_C = i_L = 9.62\text{E}{-}3\angle{-}90°$,

9. $v_R = v_C = 3.45\angle{-}80.5°$, $v_L = 7.28\angle 27.9°$

11. $i_{50} = 31.4\text{E}{-}6\angle 90°$, $i_{91} = i_{20} = 12.6\text{E}{-}6\angle 89.9°$, $i_s \approx 44\text{E}{-}6\angle 90°$

13. $v_b = 9.19\angle{-}50°$, $v_{ab} = 15.8\angle 26.5°$

15. $i_{j4k} = 5\text{E}{-}3\angle{-}90°$, $i_{2.7k} = 7.41\text{E}{-}3\angle 0°$, $i_{3.9k} = i_{-j1k} = 4.97\text{E}{-}3\angle 14.4°$, $i_s = 12.8\text{E}{-}3\angle{-}17.1°$

17. $v_b = 120\angle 0.7°$, $v_{ab} = 20\angle{-}175.6°$

19. $i_{15k} = i_L = 887\text{E}{-}6\angle{-}4°$, $i_{12k} = 1.12\text{E}{-}3\angle 3.17°$

21. $v_a = 4.8\angle 14.1°$, $v_b = 6.62\angle 14.1°$

23. $v_a = 571\text{E}{-}3\angle{-}26.5°$, $v_b = 537\text{E}{-}3\angle{-}6.68°$

25. $v_a = 19.85\angle 127.2°$, $v_b = 23.8\angle 127.9°$

27. $i_c = 12.83\text{E}{-}3\angle 136.9°$

29. $v_a = 7.32\angle 61.5°$, $v_b = 7.1\angle 75.6°$

31. $v_{ab} = {-}133.3\angle 0°$

33. 217 nF

35. 3.24 µF

37. 5.9 µF

39. 19.1 mH

5 Analysis Theorems and Techniques

1. $v_b = 2.99\angle 3.5°$

3. $i_{82} = 13.3\text{E}-3\angle 59.8°$

5. $v_b = 1.08\angle 165°$

7. $i_{2.2k} = 1.3\text{E}-3\angle 49.2°$

9. $v_b = 7.84\angle 98.9°$, $v_{cd} = 10.2\angle -44.2°$

11. $i_{S1} = 337\text{E}-6\angle -73.6°$, $i_{S2} = 401\text{E}-6\angle 152°$

13. $v_{ab} = 14.3\angle -25.4°$

15. $v_{ab} = 972\text{E}-3\angle -166°$

17. $v_a = 1.31\angle -174°$, $v_b = 2.58\angle -154°$

19. $i_C = 103\text{E}-3\angle 101°$, $i_L = 369\text{E}-3\angle 171°$

21. $v_{bc} = 19.65\angle -102.5°$

23. $i_E = 14.8\text{E}-3\angle 177°$

25. $v_b = 10.7\angle 48.4°$

27. $i_L = 286.7\text{E}-3\angle 40.6°$

29. $v_{ab} = 7.18\angle -115°$

31. $v_b = 2.68\angle 42.3°$

33. $i_{5k} = 49.9\text{E}-6\angle -29.9°$

35. $v_c = 106.7\angle 90.6°$

37. $i_{1k} = 17.8\text{E}-3\angle 21.8°$

39. $v_a = 9.7\angle 76°$, $v_b = 0.273\angle -12.3°$

41. $Z_{TH} = 91\angle 0°$, $E_{TH} = 1\angle 0°$

43. $Z_{TH} = Z_N = j714.3$, $E_{TH} = 8.571\angle 0°$, $I_N = 12\text{E}-3\angle -90°$

45. $Z_N = 514.5\angle 31°$, $I_N = 24\text{E}-3\angle -45°$

47. $v_b = 5.69\angle -71.5°$

49. $Z_N = 1564\angle 17.1°$, $I_N = 19.2\text{E}-3\angle -17.1°$, $Z_L = 1495 - j461$

51. $Z_{TH} = Z_N = 64.4\angle 18.8°$, $E_{TH} = 113.6\angle 18.8°$, $I_N = 1.765\angle 0°$, Combo needed $= 61 - j20.7$, $P = 52.9$ W

53. All three pair $= 3.33$ k $+ j3.33$ k

55. All three pair $= 1.33$ k $- j1$ k

57. $Z_a = Z_{xy} = 3.667$ k $+ j3.667$ k, $Z_b = Z_{xz} = 5.5$ k $+ j5.5$ k, $Z_c = Z_{yz} = 11$ k $+ j11$ k

59. $Z_a = Z_{xy} = 1.83$ k $- j1.83$ k, $Z_b = Z_{xz} = 2.75$ k $- j\,2.75$ k, $Z_c = Z_{yz} = 5.5$ k $- j\,5.5$ k

61. $v_{bc} = 4.5545\angle 65.4°$

63. $i_S = 11.5\text{E}-3\angle 11.3°$, in series with 4.3 k$\Omega$ and a capacitive reactance of $-j5$ k

65. $i_S = 1.54\text{E}-3\angle -22.7°$, in parallel with a series combination of 600 Ω and 2 mH

67. $e_S = 1.98\angle -49.3°$, in parallel with a series combination of 1 kΩ and 79.6 nF

69. $e_S = 96.3\angle 19°$, in parallel with a series combination of 9.1 kΩ and 5 mH

71.
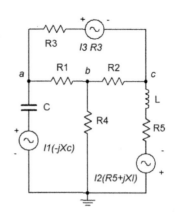

73.
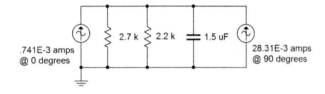

75. $-j43.8\ \Omega$

6 Mesh and Nodal Analysis

1. Loop ordering is left to right

$$2\angle0° = (4.9\text{ k})i_1 - (2.2\text{ k})i_2$$

$$-3\angle0° = -(2.2\text{ k})i_1 + (2.2\text{ k} - j106)i_2$$

3. $i_{75} = 8.22\text{E}{-}3\angle{-}138.5°$

5. Loop ordering is left to right

$$20\angle0° = (6.6\text{ k} + j4\text{ k})i_1 - (2.7\text{ k} + j4\text{ k})i_2$$

$$-5\angle0° = -(2.7\text{ k} + j4\text{ k})i_1 + (4.5\text{ k} + j3\text{ k})i_2$$

$v_b = 7.84\angle98.9°$

7. $i_{j200} = 4.36\text{E}{-}3\angle{-}28.5°$

9. $v_{cd} = 218\text{E}{-}3\angle{-}153°$

11. $i_{-j200} = 49.5\text{E}{-}3\angle{-}44.8°$ (up)

13. $v_c = 14.41\angle33.3°$

15. $i_{j300} = 11.7\text{E}{-}3\angle138°$ (up)

17. $v_{bc} = 4.545\angle65.4°$

19. $i_{R3} = 9.81\text{E}{-}3\angle175°$

21. $v_b = 9.24\angle{-}8.7°$

23. $i_{2.2k} = 1.51\text{E}{-}3\angle{-}2.04°$

25. $v_{ab} = 14.3\angle{-}25.4°$

27. $i_{330} = 7.76\text{E}{-}3\angle14.1°$

29. $v_c = 34.4\angle37.7°$

31. $i_{43} = 33.2\text{E}{-}3\angle44.4°$

33. $v_c = 180\angle{-}106°$

35. $i_{4\Omega} = 772\text{E}{-}3\angle47.1°$

37. $v_{ac} = 18.35\angle{-}80.8°$

39. $i_{22} = 3.07\angle{-}153°$

41. $v_c = 6.09\angle{-}3.68°$

43. $i_{3.3k} = 9.86\text{E}{-}3\angle18.8°$

45. $v_c = 18\angle38.9°$

47. $v_{bc} = 1.09\angle59.8°$

49. $v_{ba} = 964\text{E}{-}3\angle127.3°$

51. $v_{cb} = 15.2\angle149°$

53. $v_{bc} = 58.9\angle175°$

55. $i_{1k} = 5.24\text{E}{-}3\angle32.9°$

57. $v_{ab} = 90.6\angle{-}112°$

59. $i_{1k} = 99.95\text{E}{-}6\angle0°$

61. $v_d = 16.67\angle.191°$

7 AC Power

1. $S = 6.25$ mVA, $P = 4.88$ mW, $Q = 3.9$ mVAR (ind), PF = 0.781

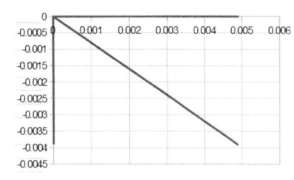

3. $S = 64.4$ VA, $P = 57.7$ W, $Q = 28.8$ VAR (ind), PF = 0.894

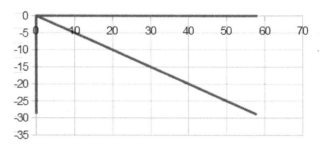

5. $S = 315$ VA, $P = 162$ W, $Q = 270$ VAR (ind), PF = 0.515

7. $S = 751$ VA, $P = 720$ W, $Q = 216$ VAR (cap), PF = 0.958

9. $i = 4.286$ A, $P = 90.9$ W

11. $P = 6.4$ kW, $Q = 4.8$ kVAR, $L = 2.86$ mH

13. $S = P = 180$ W, $Q = 0$ VAR, $i = 1.5$ A

15. $S = P = 1600$ W, $Q = 0$ VAR

17. $S = 1523$ VA, $P = 1400$ W, $Q = 600$ VAR, PF = 0.919, $i = 12.7$ A

19. $S = 1811$ VA, $P = 1800$ W, $Q = 200$ VAR, PF = 0.994, $i = 15.09$ A

21. 89.5%

23. 0.829

25. $S = 683$ VA, $P = 478$ W, $Q = 488$ VAR, $i = 5.69$ A

27. $S = 2403$ VA, $P = 2332$ W, PF = 0.971, $i = 20$ A

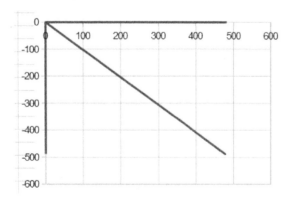

29. $S = 3154$ VA, $P = 2943$ W, $Q = 1135$ VAR, PF = 0.933

31. 141 µF

33. 102 mH

35. 17.6 mH

37. 260 μF

8 Resonance

1. $BW = 14.67$ kHz, $f_1 = 432.7$ kHz, $f_2 = 447.3$ kHz

3. $Q_{coil} = 50$, $R_{coil} = 1.88$ Ω

5. 15 kΩ $\|$ $j600$ Ω

7. $R_p = 20.36$ kΩ, $L_p = 75$ μH

9. $f_0 = 247$ kHz, $Q = 10.2$

11. $f_0 = 5.03$ kHz, $Q_{sys} = 8.35$, $BW = 602$ Hz

13. $f_0 = 10.07$ kHz, $Q_{sys} = 4.22$, $BW = 2.39$ kHz, $v_C = 4.22$

15. $f_0 = 10.07$ kHz, $Q_{sys} = 3.51$, $BW = 2.87$ kHz, $v_C = 3.51$

17. $f_0 = 10.07$ kHz, $Q_{sys} = 4.22$, $BW = 2.39$ Hz, $v_R = 1.5$, $v_C = 6.32$

19. $f_0 = 136.8$ kHz, $Q_{sys} = 11$, $BW = 12.4$ kHz, $i_L = 424\text{E}{-3}\angle{-84.8°}$, $i_C = 425.5\text{E}{-3}\angle 90°$

21. $f_0 = 360.4$ kHz, $Q_{sys} = 24.1$, $BW = 14.9$ kHz, $i_R = 3.19\text{E}{-3}\angle 0°$, $i_C = 169.8\text{E}{-3}\angle 90°$, $i_L = 169.8\text{E}{-3}\angle{-88.7°}$

23. $f_0 = 142.4$ kHz, $Q_{sys} = 14.35$, $BW = 9.92$ kHz, $i_R = 5\text{E}{-3}\angle 0°$, $i_C = 111.9\text{E}{-3}\angle 90°$, $i_L = 111.8\text{E}{-3}\angle{-88.6°}$

25. $f_0 = 225$ kHz, $Q_{sys} = 25.3$, $BW = 8.9$ kHz, $i_R = 149\text{E}{-6}\angle 0°$, $i_C = 12.6\text{E}{-3}\angle 90°$, $i_L = 12.6\text{E}{-3}\angle{-88.4°}$, $v_R = 1.787$

27. 135 μH

29. 6.94 kΩ

31. $Q_{sys} = 50$, however, none of the inductors exhibit $Q_{coil} \geq 50$ at 1 MHz. Therefore there are no viable candidates because Q_{sys} can be no larger than Q_{coil}.

9 Polyphase Power

1. $v_{LINE} = 120\angle 0°$ V (with $\angle 120°$ and $\angle 240°$, other phases not shown from here on), $i_{LINE} = 20.8$ A, $i_{LOAD} = 12$ A, $P_{LOAD} = 4320$ W

3. $v_{LINE} = 208$ V, $i_{LINE} = i_{LOAD} = 10.4$ A, $P_{LOAD} = 6490$ W

5. $v_{LINE} = 120$ V, $v_{LOAD} = 69.3$ V, $i_{LINE} = 13.86$ A, $i_{GEN\text{-}PHASE} = 8$ A, $P_{TOTAL} = 2880$ W

7. $v_{LINE} = v_{LOAD} = 208$ V, $i_{LINE} = 6$ A, $P_{LOAD} = 2160$ W

9. $v_{LINE} = 208$ V, $i_{LINE} = 33.4$ A, $S_{LOAD} = 12$ kVA, $P_{LOAD} = 11.1$ kW

11. $v_{LINE} = 120$ V, $i_{LINE} = i_{LOAD} = 11.14$ A, $S_{LOAD} = 4$ kVA, $P_{LOAD} = 3.72$ kW

13. $v_{LINE} = v_{LOAD} = 398$ V, $i_{LINE} = 13.8$ A, $i_{LOAD} = 7.97$ A, $S_{LOAD} = 9.52$ kVA, $P_{LOAD} = 7.61$ kW

15. $v_{LINE} = 120$ V, $i_{LINE} = i_{LOAD} = 11.$ A, $S_{LOAD} = 4$ kVA, $P_{LOAD} = 3.72$ kW

17. $v_{LINE} = 120$ V, $i_{LINE} = 2.747$ A, $S_{LOAD} = 571$ VA, $P_{LOAD} = 566$ W

19. 91.5 µF

10 Magnetic Circuits and Transformers

1. 80 mA

3. 95 mA

5. 160 turns

7. 20 mA

9. 2.42 A

11. Cast steel causes greater core losses so the secondary current would be reduced.

13. 324 mA

15. $v_{load} = 4$ volts, $i_{load} = 40$ mA, $i_{primary} = 6.67$ mA

17. 1440 Ω

19. 45 VA

Appendix C

Fun with Google Translate

If you never felt dumb, then you're dumb, because no one is that smart.

Si vous ne vous êtes jamais senti stupide, alors vous êtes stupide, parce que personne n'est aussi intelligent.

Wenn du dich nie dumm gefühlt hast, dann bist du dumm, weil niemand so schlau ist.

Si nunca te has sentido estúpido, entonces eres estúpido porque nadie es tan inteligente.

Если вы никогда не чувствовали себя глупо, тогда вы глупы, потому что никто не настолько умный.

Si umquam sensisse inepta stultus eris quia non est dolor.

あなたは愚かな愚かなと感じたことがない場合は痛みがないので。

Perché non c'è dolore se non ti sei mai sentito stupido stupido.

Because there's no pain if you've never felt stupid stupid.

Printed in Great Britain
by Amazon